Principles of Highway Engineering

Principles of Highway Engineering

Horton George

Willford Press
www.willfordpress.com

Published by Willford Press,
118-35 Queens Blvd., Suite 400,
Forest Hills, NY 11375, USA

ISBN: 978-1-64728-500-5

Cataloging-in-Publication Data

Principles of highway engineering / Horton George.
p. cm.
Includes bibliographical references and index.
ISBN 978-1-64728-500-5
1. Highway engineering. 2. Highway planning. 3. Roads. I. George, Horton.
TE145 .P75 2023
625.7--dc23

For information on all Willford Press publications
visit our website at www.willfordpress.com

WILLFORD PRESS

Contents

Preface

This book has been a concerted effort by a group of academicians, researchers and scientists, who have contributed their research works for the realization of the book. This book has materialized in the wake of emerging advancements and innovations in this field. Therefore, the need of the hour was to compile all the required researches and disseminate the knowledge to a broad spectrum of people comprising of students, researchers and specialists of the field.

Highway is any private or public road including other public path on land. Highway engineering refers to a branch of civil engineering that focuses on designing, building, operating, planning and maintaining tunnels, roads and bridges to enable efficient and safe transportation of goods and people. It considers aspects like future traffic capacities, highway intersection designs, geometric alignment, design and material of highway pavement, its thickness, and maintenance. The assessment of existing and future traffic capacities on a road network is part of highway planning. During the design phase, the best location, shape and alignment for a highway are chosen. This book includes some of the vital pieces of work being conducted across the world, on various topics related to highway engineering. It is appropriate for students seeking detailed information in this area as well as for experts.

At the end of the preface, I would like to thank the authors for their brilliant chapters and the publisher for guiding us all-through the making of the book till its final stage. Also, I would like to thank my family for providing the support and encouragement throughout my academic career and research projects.

Horton George

1

Highway Planning and Alignment

1.1 Significance of Highway Planning

For development of highways, a proper highway planning is needed.

Every highway planning is done keeping in mind the following significance:

- To establish an integrated highway network with minimum cost capable of encompassing all highway travel in a most safe, efficient, economical and orderly manner. The minimum cost should include initial cost of construction, maintenance cost, renewal cost of pavement layers and the vehicle operation costs.
- To plan a highway system, which could be constructed within the available resources during the proposed plan period with maximum utility.
- To forecast the future requirements of road kilometer needed and improvement of roads.
- To phase out the entire highway development during the plan period by identifying the priorities keeping in view, the utility as the main criterion.
- To work out financing system.

1.2 Modal Limitations towards Sustainability

Sustainable Development

Sustainable development is a complex concept that is subjected to numerous interpretations. It is generally agreed that a sustainable society favour conditions that benefits the environment, the economy and the society without compromising the welfare of future generations.

The conditions of the future societies will largely depend upon the legacy of current societies on resources and the environment. All form of assets passed on to the next generation should be at least of equal value per capita.

The basic definition of sustainability has been expanded to include three major points:

1. Social Equity

It relates to conditions favouring a distribution of resources among the current generation based upon comparative levels of productivity. This implies that the individuals or the institutions are free to pursue the ventures of their choice and reaps the rewards for the risk they take and the efforts they make.

It should not be confused with welfare programs where the productive segment of the population agrees or is coerced to support a non-productive segment.

2. Economic Efficiency

It is concerned with conditions permitting higher levels of economic efficiency in terms of resource and labour usage. It focuses on capabilities, competitiveness, flexibility in production and providing goods and services that supply a market demand. Under such circumstances factors of the production should be freely allocated and markets open to trade.

3. Environmental Responsibility

It involves a "footprint" which is lesser than the capacity of the environment to accommodate. This includes the supply of resources but also the safe disposal of numerous forms of wastes.

Its core tenets include the conservation and reuse of resources. Societies do not contribute to environmental problems at the same level.

A comparison between developed countries and developing countries reveals that the developed world consumes 70% of the world's energy, 75% of minerals and 85% of wood.

Sustainability can be thus expressed at two spatial levels:

Global

It includes long term stability of the earth's environment and availability of resources to support human activities. Localized forms are often related to urban areas in terms of jobs, housing and environmental pollution.

Since a growing share of the global population is urbanized, sustainability is increasingly focusing on urban areas. Major cities require a vast array of supporting infrastructures including the energy, water, sewers and transport.

A key to urban sustainability issues are linked with the provision and maintenance of a wide range of urban infrastructure. Every city has specific infrastructure and environmental problems. For instance, cities in developing countries have chronic deficiencies in the provision of the most basic infrastructure while their environmental conditions are deteriorating. Infrastructures can be publicly or privately owned.

The public infrastructures have the advantage to be available to a larger share of the population at a low cost but are expensive for government to maintain subsidies.

Private infrastructures tend to service a smaller share of the population, at the choice of Infrastructure Company but are financially profitable. As income levels increase, some infrastructure problems are solved while some environmental problems are created.

For instance, an increase in income is linked to better sanitation and water provision, but at the expense of greater waste and carbon dioxide emissions.

Sustainable Transportation

Transportation as a core component supporting the interactions and the development of socioeconomic systems has also been the object of much consideration about to what extent it is sustainable.

Sustainable transportation can be defined as:

The capacity to support mobility needs of people, freight and information in a manner that is the least damageable to the environment. Sustainable development applied to transport systems requires the promotion of linkages between the environmental protection, economic efficiency and social progress.

Under the environmental dimension, the objective consists of understanding the reciprocal influences of the physical environment and the practices of the industry and that the environmental issues are addressed by all aspects of the transport industry.

Transport must be cost-effective and capable of adapting to changing demands. Under the social dimension, the objective consists of upgrading standards of living and quality of life. Most agree that automobile dependence is related to an unsustainable urban environment.

But the mobility choice and preferences of the global population where the automobile is rapidly adopted when income levels reach a certain threshold. Other transport alternatives commonly do not measure up to the convenience of the automobile.

Private and flexible forms of the transportation such as the automobile are thus fundamental to urban mobility and should not be discarded as options for the sake of sustainability. A bias is observed in the transport community towards an emphasis for the public transit and non-motorized transportation as the dominant.

Almost all public transit systems are financially unsustainable imposing burdens on the society. The freight transportation must also be involved in this process considering the substantial growth of raw materials and goods being traded in a global economy.

Freight transportation is relies on much more environmentally sound modes such as rail and maritime transport. Despite the apparent and projected success of measures to promote transport sustainability they have their limits.

Built environment and transport infrastructure cannot change quickly enough to solve the bulk of problems related to unsustainable transport. The transportation and the sustainability for both passengers and freight must also contend with mitigation versus adaptation issues.

The mitigation concerns the improvement of productivity and efficiency of existing modes, terminals and managerial approaches so that environmental externalities are reduced. They tend to be short to medium term strategies.

Adaptation is a change in the level of use and the market share of respective modes to better reflect a long term trend such as higher energy prices and stricter environmental regulations. There is a wide range of responses to environmental sustainability.

The various outcomes for the sustainable environment involve three steps:

- Transport operations must conform to local, national and international regulations.
- Environmental costs of transport operations must be built into the price of providing transport facilities and services.
- The environmental performance must be introduced into the organization's management.

The environmental sustainability represents a growing area of responsibility for transport companies, one that is forcing them to acquire expertise in environmental management.

The most important challenge for the industry is to implement environmentally sustainable transport within competitive market structures leaning on coping with changes in transport demand while improving transport supply.

Managing Transport Demand

In order to effectively mitigate the adverse impacts of current transportation systems, strategies must be devised to manage transport demand for passengers and freight as well as to redistribute this demand in space or in time when possible.

If unsubsidized transportation is inexpensive it is a good signal of its sustainability. Increasing transport costs and the pressure to subsidize them can be interpreted as signals that they may be unsustainable.

There are several interrelated ways in which transportation systems can adapt to cope with the transport demand and reach a better level of sustainability.

The densification and agglomeration of activities can involve spatial structures such as the logistics zones or transit oriented developments and can result in reduced vehicle trips and increase some of sustainable transport alternatives.

It may incite the use of modes more prone to economies of scale such as public transport and pedestrian or bicycle travel as cost-effective alternatives. With strong market signals related to the land cost, land economics often dictates more efficient and higher density uses.

Full-cost pricing involves the full or partial recovery of costs related to the public investments incurred by varying levels of government in relation to the constructing, repairing and operating transport networks.

They remove artificial signals such as subsidies and let users assume the real cost of transportation. It includes road pricing, pollution taxes, fees and parking controls.

Motorists are charged a floating fee depending on the variability of demand in peak and off peak hours for using the targeted roads. This can be implemented through a variety of techniques such as tolls or licensing fees.

Parking controls either in raising parking prices or reducing the amount of parking areas can be used to determine the use of privately owned vehicles in areas of highest demand by raising the price of commuting to high density areas.

The expected result of this strategy is to encourage commuters to seek cheaper alternatives either in mass transit or carpooling.

The strategy of traffic bans is a more direct method of reducing traffic demand. Through traffic bans, the responsible regional or the municipal institutions would exert direct control over the allowable limit of vehicles in a given urban area depending on measures of transport supply-demand functions or arbitrary estimates of carrying capacity.

The implementation of such strategies relies heavily on the existing spatial structure, people, material flows and transport networks. An expectation is that the demand will shift towards modes that are more environmentally efficient and having a better energy performance.

In situations where a fee structure is not effective (e.g. low income population), constraint based strategies can be more effective than the fee based strategies.

Such coercive strategies would thereby impose a limit on the number of vehicles in circulation and correspondingly reduce congestion and air pollution while promoting the use of alternative means of transport.

Improving Transport Supply

The implementation of demand oriented policies and mechanisms are an important component in promoting sustainable transport. It is pivotal that these measures be coupled with transport supply improvements including improving vehicle propulsion technologies.

Transportation infrastructure should be expanded to accommodate rapidly growing transport demands. As the global urban population continues to grow particularly in developing countries there are pressures to expand urban transport infrastructures.

In urban areas challenge is to expand and improve the transportation supply in such a way that the automobile and trucking are only part of the equation rather than the focus.

In order to implement effective transport supply expansion policies, fast, efficient, reliable and accessible transport alternatives must be provided preferably through private initiatives.

For passengers this can be achieved by expanding public the transit infrastructure, by improving existing public transit services and by making cities friendly to the pedestrians and non-motorized vehicles.

The issue of sustainability is giving public transit a new impetus since the bulk of its prior rational was to mitigate automobile dependency and provide equity in mobility.

Automobile dependency is the outcome of market forces expressed as consumer preferences and national manufacturing policies. Thus, alternatives can be provided if they prove to be cost effective while fulfilling a niche demand. They may include:

- Reconciling public and private transport.
- Both forms of transportation have roles to fulfill.

The long term trend has been a substantial growth of private transport in spite of all the efforts made to promote public transportation.

The expansion and development of the mass transit systems must not only satisfy a niche demand but also make effective use of urban space by conforming to a number of factors including urban form, density and modal preferences.

In doing so, the fleets and networks must ensure a level of flexibility while ensuring low ridership costs which would reduce per-passenger pollution levels and traffic congestion alike. Comparatively methods of improving and upgrading existing public transit services should include the improvement of service coverage and quality as well as increasing frequency where and when it is most needed.

Also the chosen modes of mass transit transportation whether bus, light rail, subway, trams, group taxis or a combination of two or more should be designed, implemented and operated in a way that will facilitate and predicate complementary and interchangeability while remaining cost-effective.

The integration of individual modes of non-motorized transport such as walking and cycling if properly promoted and encouraged can provide access to shopping, schools and work for a growing proportion of the population.

For cities struggling with serious traffic congestion and air pollution, non-motorized transport should be considered as an indispensable and beneficial alternative to private vehicles while serving as a crucial link in an integrated public transportation system.

Such alternatives are however in contrast with the modal choice reality of many economies towards the automobile particularly those experiencing rapid growth. The issue of sustainable transport thus remains elusive.

1.3 History of Road Development in India

Existence of roads in India dates back to the remote past. Highway Development in India from ancient age to modern age is briefly discussed below:

Before B.C.

Excavations of cities like Mohanjodaro and Harappa which existed between 3500 BC to 2500 BC have revealed the existence of broad streets. It is found that the importance of roads was realized particularly for military and administrative purposes.

Ancient scripts reveal that a road of 6 to 7.5 m wide was built by King Bimbisara in 600 BC. Ancient literature known as Arthashastra furnishes adequate information on the road construction during the Aryan period.

Kautilya, the first Prime Minister of Chandra Gupta Maurya constructed a highway connecting North West Frontier province to Patna during 300 BC. During 270 BC King Ashoka constructed a good network of roads.

Mughal Period

Road development received special attention during this period. In this period twenty four long roads connecting different parts of the country were constructed. Out of which thirteen were traced. During this period roads were greatly improved and well maintained. Mughals received great appreciation from the foreign visitors who visited India during the region.

British Period

At the beginning of British rule, the conditions of roads in India were very much damaged. Economic and political shifts caused deterioration to road transportation. The fall of Mughal's rule led to the ruin of the road communication system.

Due to military movement, Britishers got a few truck roads metalled and got bridges constructed. Only old roads were repaired or metalled. Lord Dalhousie, the Governor General of India formed the Central Public Works Department to look after the roads in India.

The construction of Grand Trunk Road was undertaken by this new Department. Lord William Bentinck, Lord Mayo and Lord Ripon contributed measurably to the road construction in India. The introduction of railways in 1853 was a serious setback faced by road construction.

Only a few road constructions of military importance were carried out. After World War I, a revolution in India's transport system came due to the cheap supply of surplus military vehicles.

In order to bridge the traffic due to motor vehicles and village bullocks, refinements were needed in the practice of road construction. Under this condition Jayakar Committee was formed, recommendations of the committee were implemented in phased manner.

During the World War II (1939), the roads were under the control of provinces and due to lack of funds, the provinces could not meet the requirements of road development. After the war period, it was realized that the effective system of roads could be active only if it is under central government. It was felt if not all roads at least a few of them could be maintained by the central government.

A conference of chief engineers was held at Nagpur in 1943 which made very useful recommendations. Conference has been commonly known as Nagpur Plan.

Jayakar Committee Recommendations and Realizations

After the World War I, number of motor vehicles gradually increased and demanded a better network of roads. There was a need for the roads to carry both bullock cart traffic and motor vehicles.

But existing roads were not laid to withstand the mixed traffic conditions. In 1927, the Indian Roads and Transport Development Association (IRTDA) were set up having head office at Bombay and branch offices at Delhi, Gauhati and Madras (Chennai).

Its function was to study the transport problems of the country and recommend improvements. Mr. M.R. Jayakar was appointed as the Chairman by the Central Government for a committee called as "Jayakar Committee".

The committee looked into the road development in India and submitted a report based on the recommendations of IRTDA. The Jayakar Committee submitted its report in 1928.

Important Recommendations of Jayakar Committee

Road development should be considered as a national problem and necessary attention should be paid. To meet the road development expenses, an extra tax to be levied on petrol and to be named as central road fund.

An advisory board on road development has to be constituted with experts pooled from different sectors who have knowledge on road development. A research organization to be set up at national level to undertake research on roads and road related areas and provide consultancy services to the Government as and when needed.

As per the recommendations, "Central Road Fund" was created on 18th March 1929. The consumers of petrol were then charged an extra levy of 2.04 paise per litre (i.e., two annas per gallon).

Twenty percent of the revenue collected through the fund was retained as central reserve and the balance allotted to the various states based on actual petrol consumption. Based on the recommendation, a semi-official technical committee known as Indian Roads Congress (IRC) was formed in 1934.

Major Roles of IRC

- To promote and encourage the science and practice of construction and maintenance of roads.
- To suggest improved methods of administration, planning, design, construction, operation, use and maintenance of roads.
- To promote the use of standard specifications and to propose specifications.
- To advise regarding education, experience and research connected with roads.
- To suggest legislation for the development, improvement and protection of roads.
- To publish or arrange for the publication of proceedings, journals, periodicals and other literature for the promotion of the objects of the society.
- Based on the recommendations of the Jayakar Committee in the year 1950, a research institute exclusively for research on roads was started under the name, Central Road Research Institute (CRRI) at New Delhi.

Functions of CRRI

Central Road Research Institute is engaged in carrying out the following functions:

- Research and Development projects on design, construction and maintenance of roads.

- Traffic and Transportation planning of mega and medium cities.
- Management of roads in different terrains.
- Improvement of marginal materials and utilization of industrial waste in road construction.
- Landslide control and ground improvements.
- Environmental Pollution.
- Road Traffic safety analysis and design.
- Service life assessment and rehabilitation of highway and railway bridges.
- Performance monitoring and evaluation of roads.
- Technical and consultancy services to various user organisations.

1.4 Classification of Highways

Roads are generally classified based on the following factors:

- Traffic volume.
- Load transported or tonnage.
- Location and function.
- Speed of vehicles.

Classification based on traffic volume or tonnage have been fixed by different agencies arbitrarily hence it may not be agreeable to all.

Based on traffic volume, roads may be classified as heavy, medium and light traffic roads. These terms can be expressed as number of vehicles per day.

Load or tonnage is also relative and roads may be classified as:

Class I, Class II, Class III, etc. and the limits may be expressed as tonnes per day.

The classification based on location and function has been accepted by everyone.

Accordingly the Nagpur Road Plan classified the roads in India into five categories, viz:

- National Highways (NH).

- State Highways (SH).
- Major District Roads (MDR).
- Other District Roads ODR).
- Village Roads (VR).

These classifications have been accepted fully and have been functioning over the years. The classification based on speed of a vehicle is new one which is not fully defined.

However, roads identified to allow vehicles to go with a minimum speed have been classified as expressways.

National Highways

These are the main highways running through the length and breadth of the country.

It connects ports, foreign highways, capitals of large states and including roads required for strategic movements for the defense of India.

For underdeveloped areas it was proposed to construct national traits so that at a later stage they could be connected into appropriate roads to suit the traffic requirements.

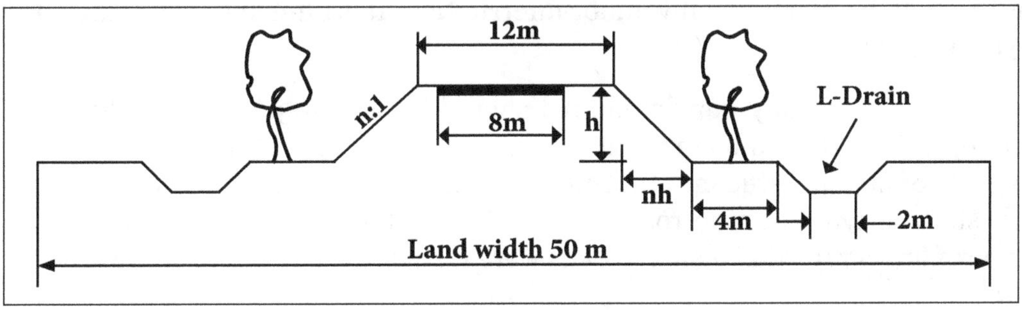

Typical cross section of National Highway.

National highways should form the framework of roads on which the entire road construction of the country should be based. Selection of National highways should be such that they afford uninterrupted road communication.

It is not mandatory that all National Highways should be same type and dimensions. All the National Highways are assigned a specific number.

Example - Highway connecting Delhi-Chandigarh-Amristar is denoted as NH-1.

Construction and maintenance of the National Highways rest with the Central Government Departments such as Central Public Works Department.

Typical cross section of a National Highway is shown in figure. National Highways are of two lanes of at least 8 m width and at least 2 m wide on either side.

Provincial or State Highways

These are all other main trunk or arterial roads of a province or state, connecting up with the National Highways of adjacent states, district headquarters and important cities within the state. It serves as the main arteries of traffic to and from district roads.

These are main roads of commercial interest to move goods within a state.

They may sometimes allow heavier traffic than National Highways and for this sake its designation cannot be changed.

These roads are of two lanes and have similar cross section as that of National Highways. Construction and maintenance are handled by State Government Public Works Department. Central Government grants are provided for the development of State Highways.

District Roads

These are those roads which are traversing each district serving areas of production and markets. It connects the district serving areas of production and markets with each other or with highways and railways. They take road traffic into the main parts of rural areas through the year with only minor interruption. It is built and maintained by local self-government.

Every village in a highly populated area should be within 3.2 km of a district road. These roads should form main branches from the National or State Highways, should be capable of taking traffic into the interior of each district or similar unit of the area. Figure shows a typical urban road of a district. With the view of expected traffic and the type of finish provided district Roads were further divided into two classes as given below.

Major District Roads

It connects all important towns of a district with headquarters. Also connects the areas of production, market, railway stations and State Highways. They roughly have the same specification as that of State Highways.

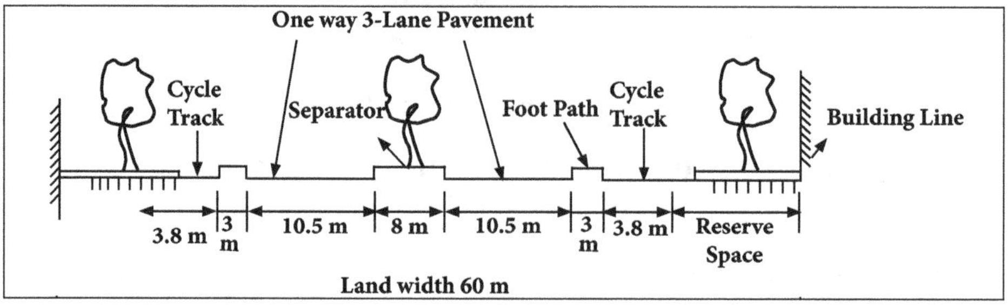

Typical Cross Section of Urban Road.

Other District Roads

These are also called as Minor District Roads. They connect a town with other towns or important villages of a district. They also connect other roads with State Highways. These are of somewhat lower specifications than the above types. During the monsoon periods many of these roads are kept closed. To be used throughout the year, necessary drains should be made. These are built and maintained by Municipal Boards and Corporations.

Village Roads

These are the roads which are connecting villages and groups of villages with each other and the nearest district road, main highway, railway or river ghat. These roads are unmetalled, connecting villages together or a village to a railway station or district road.

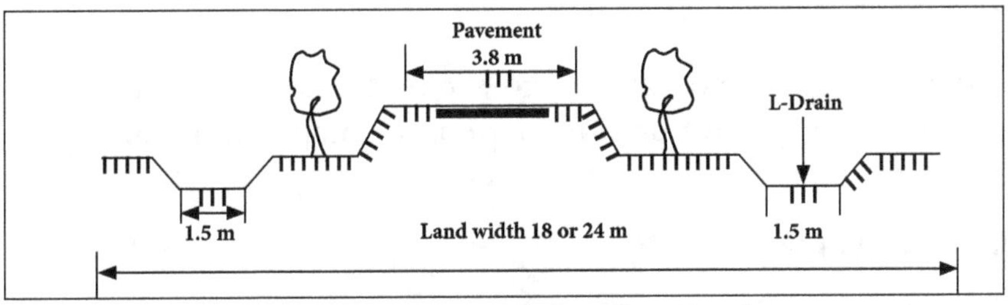

Typical Cross Section of a Village Road.

The village tracks should not be left to develop by itself, but should be replaced by a plan for a designed and regulated system. Local District Boards are responsible for the construction and maintenance of village roads. Typical cross section of a village road is shown in figure.

1.5 Locations and Functions

The aim of location survey is to select a route with the following points kept in mind:

- With reasonable economy it should meet the minimum requirement regarding curvature and grades.
- To produce an easy travelling, free flowing traffic artery that has a high capacity and it meets all the safety standards.

The location survey should recognize and evaluate the routes impact on already existing industries, business and residential values and on future development. Before field survey for any highway location is started, tentative decision regarding the design speed of the route, its cross section and the maximum grade must be made.

Decisions made are based on the estimated amount, character and hourly distribution of traffic along with knowledge of the area traversed and available funds.

After comparison of the different alternative design, the route is selected which cheapest overall has cost considering capital investment, maintenance, expense and saving to the road user.

The Rural Highway Location and Design Process

The possible desk study for highway of an area include maps, aerial photographs, charts or graphs to obtains engineering data, environmental data, social data and economic data.

Engineering Data

All details of the topography, soil and problems like drainage and maintenance should be investigated before a scientific plan programmed can be suggested.

Engineering data includes topographic and geological maps, stream and drainage basin maps, climatic records, and preliminary survey maps of previous projects, traffic surveys and capacity studies.

Environmental Data

It includes the following:

- Agricultural soil surveys indicating soil erodability.
- Air pollution studies.
- Noise and noise attenuation studies.
- Fish and wild life inventories.
- Historical studies.

Social Data

The social data obtained may be analyzed for future trends in development of an area. It includes:

- Demographic and land use information.
- Census data.
- Zoning plans and trends.
- Building permit records, motor vehicle registration records.
- Living standards of locality.

Economic Data

The economic data is essential to study the various financial aspects like source of income and manner in which funds for project may be mobilized. It includes overall cost of previous projects; unit construction costs data, agricultural, economical and industrial data etc.

1.6 Factors Influencing Highway Alignment

An alignment should be straight and short. Though not practically feasible due to various practical difficulties it has to be satisfied along the route. A short route may have a very steep gradient which cannot be accepted as the vehicles have to spend more fuel and at the same time the vehicle may undergo more strain.

Short route may face some construction and maintenance problems. Shortest routes are deviated to satisfy certain obligatory factors. Sometimes long straight routes on a flat terrain are monotonous to drivers. Hence after a few kilometers it is desirable to have a slight bend to break the monotony and keep the driver alert.

A road which is economical now need not continue to be economical in maintenance or the vehicle operations cost. A straight and short alignment may not be economical due to the ground and weather conditions.

Hence an alignment may not satisfy all the requirements and factors but has to exercise a proper judgment to decide considering all the factors. The following factors such as technical, obligatory, economy and other considerations control the highway alignment.

They are:

- Traffic,
- Geometric Design,
- Obligatory factors,
- Economy,
- Other considerations.

Traffic

It is the prime factor to be considered. It comprises of type of vehicle it has to cater and the volume of traffic it has to handle. Volume of traffic is an important factor that specifies the quantity of movement per unit of time at a specified location.

Origin of the volume of traffic and destination study should be carried out. On the stipulated area, the desired lines are considered along with the traffic flow. Proposed new road alignment should also take into account the traffic flow patterns.

Future plan as per the National road plan should also be considered e.g. National Highway, Major District Highway etc.

Geometric Design

It is the other technical factor which should be given due considerations. Geometric design factors such as gradient, radius of curvature, formation width, carriage way, right of way, sight distance, etc. would be considered in the alignment of the road.

In general the gradient to be less than the ruling gradient or design gradient for a smooth movement of vehicles. Extra widening of the road along the curve, the super-elevation to be provided, speed of vehicle, etc. are to be studied after fixing an alignment.

If the alignment is not satisfactory for the above geometric design factors, the alignment may have to be changed. Studies are to be made at horizontal curve locations and vertical curve locations to find whether the alignment will be satisfactory or not.

Line of sight, sight distance and stopping sight distance which will be available at every section should be studied. Overtaking operations of vehicles should be smooth for the permissible speeds of the vehicles on the road at the selected alignment.

Obligatory Factors

Apart from the general requirements certain requirements are mandatory or control points for the alignment of a highway.

They may be divided broadly into two categories viz:

- Points through which the alignment should pass.
- Points through which the alignment should not pass.

Obligatory Positive Factors: In order to satisfy the obligatory points through which the road alignment should pass, it may cause the alignment to deviate from the shortest or the economical path.

Following are some of the obligatory points through which or near to the alignment should pass:

Hill or Ridge: When the alignment has to cross a hill or a ridge, the alternatives are:

- To cut a tunnel.
- To go round the hill.

- To deviate fully till a suitable hill pass is available.

Figure shows the deviation of alignment AB to ACB and thereby avoiding a tunnel or large cutting:

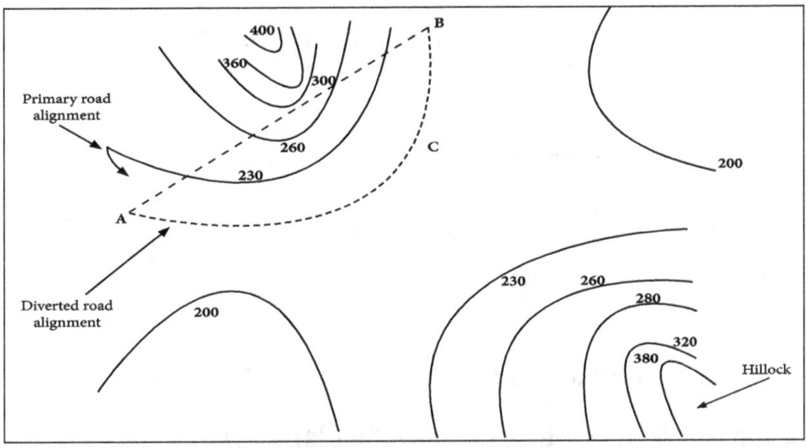

Alignment along a hill side.

River Crossing: Here the road has to cross in general perpendicular to the river center line at that point. Only at the location, bridge can be constructed. If such a point is not available oblique bridge alignment is made. For sight distance and line of sight considerations, perpendicular straight intersections are preferred.

The bridge location should have good foundation so as to be economical. Hence a straight alignment AB has to be taken along ACB. Alterations are also needed in case of railway, irrigation and navigation canals.

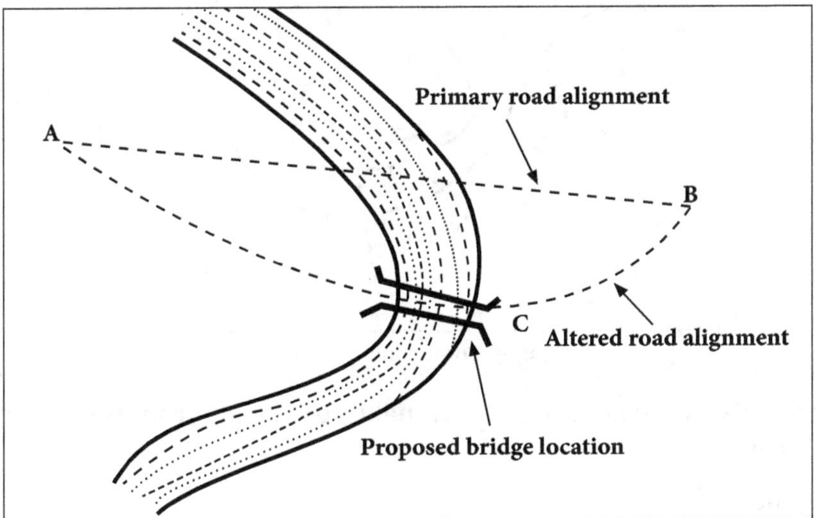

Alignment to suit proper location of bridge.

Connecting to Stations: The main aim of road alignment is to have short path, sometimes it may be necessary to connect some of the towns and villages to have better service of the road. In such a case, the road has to be diverted to suit the situation. Thus the straight alignment AB is diverted through a town C which is being linked to two villages D and E.

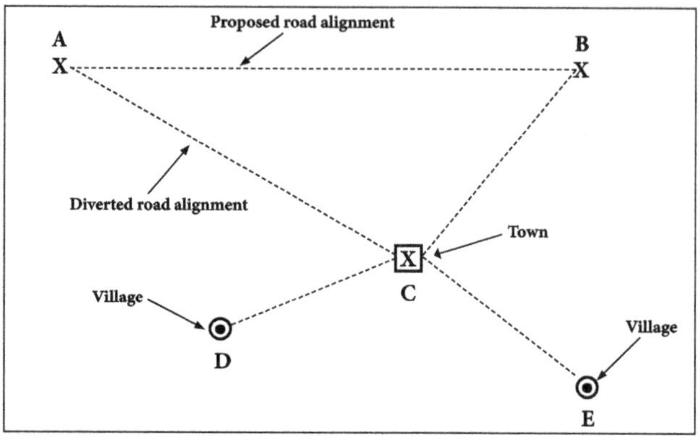

Alignment connecting intermediate towns and linking villages.

Such deviations are to be made to give a spectacular view of a canyon for a scenic park and connect an agricultural land or a farm road.

Obligatory Negative Factors:

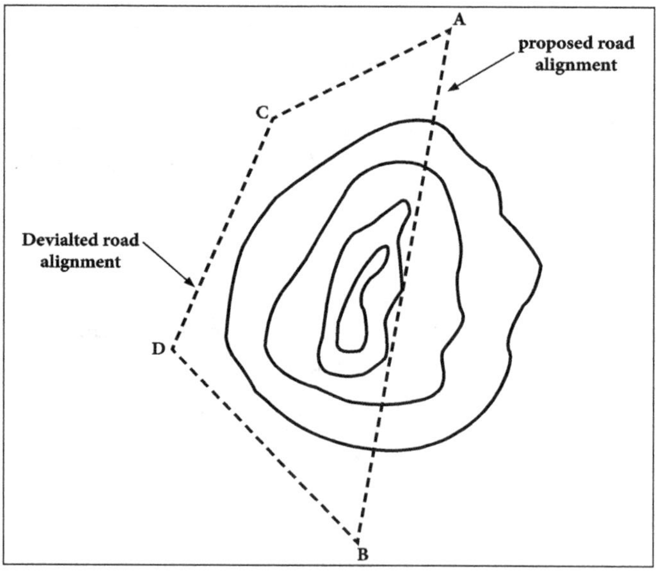

Alignment deviating from a lake.

Conditions may also arise in practice that makes to deviate from short route. Such situations are discussed below:

- Religious Places.
- Monuments.
- Unsuitable Land.

- Large Water-body.
- Agricultural Land.

Economy

The two technical and obligatory factors considered above should be basically economical. While working out the economic aspects of the road alignment, the initial cost, the maintenance cost and vehicle operations cost should be taken into account. Initial cost of construction may be reduced by slightly altering the alignment such that construction of embankment or heavy cutting may be avoided or at least reduced. A balanced alignment may be made by equalizing the cost of cutting and construction of embankment.

Other Considerations

Factors like drainage considerations, hydrological factors, monotony, etc. also govern the final alignment. Drainage considerations, sub soil conditions, cross-drainage works, crossings may also affect the alignment. Border roads may pose special problems according to the relations with the neighboring countries.

1.7 Soil Suitability Analysis

Soil survey is carried out by obtaining soil samples 1.5 to 3.0 m below the ground level or the finished grade line. In locations where embankments are planned, sampling should be done to a depth equal to twice the height of the finished embankment. Spacing of sampling and type of sampling depends on soil type at the location.

During survey one may cross areas where land is loose or is subject to slides and there may be stretches of rocky strata, which has to be recorded. Efforts are to be made to avoid loose soils. Rocky strata though costing very much in the initial stage of construction may be preferred.

Construction materials needed for highway construction are soil and stones. For the formations of highways, the primary operation involves excavation and embankments. To get a general idea of the soil all along the center line, one has to take samples. Formation of a pavement involves construction of an embankment, undulated ground cutting.

The borrowed soil for making embankment is obtained from the adjoining areas of the highway or within the highway land itself. Hence additional road land may be acquired temporarily for borrowed road material. After completion of the project the additional road area may be handed over to the owners. This arrangement may work out cheaper than owning permanent land or getting material from far of places.

1.8 Road Ecology

It focuses on the relationship of the road network, the largest human artifact on the planet and the land on which the roads are built. Road network is a net that humans have laid over the land. The roads tie the land together for humans but separate nature into pieces.

At the same time, nature degrades roads necessitating constant maintenance to keep the network usable. Road ecology incorporates elements and contributes knowledge to many other scientific disciplines. The principles of landscape ecology form the theoretical basis for most studies within road ecology. Road ecology as a science also borrows heavily from conservation biology, fish and wildlife biology.

Road ecology is an applied science and much of the work within this field endeavors to find ways to minimize the detrimental effects that road systems can have on plant and animal populations, air and water quality and human communities.

The body of knowledge generated by studies of the ecological and social effects of road systems also informs many of the current visions of sustainable development. Other highly visible outputs of road ecology include advances in the management of storm water runoff, transportation and land-use planning and the development of crossing structures that allow animal's safe passage across busy roads.

Road ecology (The Loing Viaduct, France).

Elements of Road Ecology

The science of road ecology is concerned with understanding how roads affect ecological processes, often with the goal of developing strategies for mitigating any negative effects that roads may have on the environment. Road ecology at present is focused primarily on documenting and predicting the effects of roads on plants and animals and in understanding how roads affect the movement of water and sediments.

1.9 Engineering Surveys for Alignment: Conventional and Modern Methods

Conventional Engineering Surveys for Alignment

Engineering surveys are to be carried out before finalizing any highway alignment. The first step in highway alignment and design is to procure all available pertinent data. Engineering surveys for alignment comprise of four stages. The four stages are:

- Map study.
- Reconnaissance survey.
- Preliminary survey.
- Final location survey.

Of the above four stages, first three stages of study and survey are conducted to arrive at alternate alignments keeping in view the salient features. The fourth stage of survey goes into conducting a detailed survey on the selected alignment.

Map Study

Different form of maps, aerial photographs, charts or graphs, etc., are collected. Topographical maps prepared by Survey of India, Government of India, are available with 15 or 30 meter contour intervals. They also provide the details of rivers, hills, valleys.

Careful study of the maps at the office, gives the basic details to fix serious alternate alignments. Details of these alignments may be obtained on the site. Possible alignments can be decided based on the following details available on the map. The details are:

- Alignment bypassing valleys, lakes or ponds.
- Availability of a mountain passes in a mountain range.
- Deciding the location of bridges for crossing rivers avoiding de-touring of road.
- Connecting two station points at different elevations by adopting a suitable permissible or ruling gratdient.
- Computation of length of roadway using the contour interval and the ruling gradient.

Basic study of topographic maps enables to select alternate routes which are feasible to proceed. This basic selection of route should be further studied in the field.

Reconnaissance Survey

Reconnaissance survey may be aerial reconnaissance or ground reconnaissance.

Aerial Reconnaissance

Provides a bird's eye view of the alignments under consideration. It also gives the details of surroundings along the routes. Detailed study of the reconnaissance may enable for change of alignments based on the actual features seen. A final selection may be made by a close view of the selected alignments.

Ground Reconnaissance

It consists of general examination of the ground by walking or riding along the selected routes.

The fixed survey party inspect in parallel different routes based on the alignments proposed on the basis of map study.

Each group collects all available information necessary for evaluating the merits and demerits of each proposed alignment. In order to identify the reference point's pegs should be driven at different locations.

Simple instruments like Abney level, Tangent clinometer, Pedometer, Aneroid barometer, Ghat tracer, etc., are carried by the survey party. The details to be collected during ground reconnaissance are given below:

Topography: It includes plain, rolling, steep valleys, ponds, lakes, marshy land, cliffs, gorges, etc.

Geometries: It includes gradients, radius of curves, hair-pin bends, right of ways, constraints like monuments.

Roads: It includes types of roads such as jeep track, earthen track, length along each alignment, existing conditions, elevations, etc.

Soil Type: It includes nature of soil, marshy land, flooding area, sand dunes, black - cotton soils, field identification, geological features, etc.

Climatic Conditions: It includes monthly maximum and minimum temperatures, rainfall, average annual peak period, wind direction and velocities, fog conditions, sunny days, snow fall, etc.

Crossings: It includes cross-drainage structures, railway crossings, bridge and culverts, maximum flood level, ground water level, water table fluctuation, existing highway crossings, etc.

Resource: It includes construction materials, labour-availability, locations of stone quarries availability of water, communication facilities, etc.

Economic Factors: It includes population, marketing centre, land value, agricultural and economic potential of area, environmental factors, etc.

Obligatory Points: Ancient monuments, burial grounds, religious structures, hospitals, schools, location of towns and villages, agricultural lands, industrial areas, etc., Ground reconnaissance may be followed by an aerial reconnaissance if the area is vast. The alignments proposed based on the map study may be altered on the basis of the reconnaissance. A few of the alternate alignments may be chosen for further study based on the details available from the reconnaissance survey.

Preliminary Survey

Preliminary survey is the important survey which decides the final location.

This is conducted with the following objectives:

- The proposed alignments based on reconnaissance survey are accepted at this stage and all the necessary details of topography, soil and drainage details are collected.
- The proposed alignments with the background details are checked with general requirements needed for an ideal alignment.
- The details of construction cost of every proposed alignment are made.
- The best alignment from all considerations is finalized.

This survey is conducted on the finalized alignment with large scale instruments.

It may be carried by following conventional approach or modern approach depending on the magnitude of the project. The conventional approach has all the following components, viz., Primary traverse, Topographical features, Levelling work, and Hydrological data and Soil survey. Modern rapid approach is by conducting aerial survey by adopting aerial photographs.

Conventional Preliminary Survey

It is discussed in the following paragraphs:

Primary Traverse: An open traverse is conducted along the alignments. The length of center line should be measured by using chaining methods more accurately.

Instead a tachometric survey may be conducted. For alternate alignments a secondary traverse or an independent primary traverse may be necessary.

Topographical Features: In the next set up, the topographical features are recorded soon after establishing the center lines by the primary traverse.

Details along the center line such as geographical and other manmade features on either side of the central line are surveyed and plotted.

Width of survey to be conducted is the minimum land width or as decided by the surveying party.

Levelling Work: Levelling work is also to be carried out to get the central line profiles and typical cross sections. A minimum leveling work is done along the alternate alignments to obtain the approximate earthwork.

Hydrological Data: The next step is to decide the type, number and approximate size of cross drainage structures. For this necessary hydrological data are collected.

For the vertical alignment, the grade line is decided considering the hydrological and drainage data.

Soil Survey: Decision of the alignment depends on the type of soil on which the maximum length of road alignment passes.

This survey conducted at this stage is helpful for the following activities:

- To work out the details of earthwork.
- To find the suitability of materials for construction.
- To check the slope, gradients and their stability.
- To decide the type of pavement.
- To decide the base and sub-base courses.
- To find the surface drainage requirements etc.

To get the soil details, a detailed soil sampling need not be done at this stage but auger boring may be enough.

Samples all along the alternate alignments may be taken to get the field identification followed by laboratory basic tests to get preliminary idea of the soil profile. Geophysical and electrical resistivity methods may yield the location of hard stratum and thickness of other layers. Soil samples collected on the soil survey are used to get the soil consistency and classification. Soil profile is drawn roughly all along the alternate alignments up to the depth of exploration.

Aerial Preliminary Survey

This is conducted if the area to be surveyed is vast by following the steps given below:

- Photographs of the strip of land are taken with the required longitudinal and lateral overlaps. In order to prepare mosaic, vertical photographs are needed.

- The photographs are examined under stereoscopes and control points are selected. These are used for establishing the traverse of the alternate alignments proposed. The control points are located on the maps.
- Spot levels and subsequently contour lines may be obtained using stereo-pair observations. Also the topographic details may be noted down on the maps.
- Photo-interpretation techniques are used to decide the geological features, soil conditions, drainage requirements, etc.

With the data collected from the preliminary survey of all the alternate alignments, a comparison is made. The final selection of alignment is made keeping in view the requirements and the factors to be satisfied for the best selection of final alignment. Final location and detailed survey is conducted on this selected alignment.

Final Location Survey

1. Design Office Work

After the conclusion of preliminary survey, plans and longitudinal sections are prepared so as to go ahead for detailed location survey. Sharp curves, hair-pin bends, bridge crossings, etc. are shown on the plan. Contours at 1 to 3 m intervals are drawn to facilitate the final decision.

Scales for the map be adopted as 1:1000 horizontal and 1:100 vertical. As for as possible the same scale is adopted in the final drawing also. On certain locations, e.g., at hair pin bends, steep terrain, etc., plans are required to be drawn on a smaller scale.

In general, for built up areas and hilly terrain a horizontal scale of 1 in 2500 for horizontal scale and 1 in 250 for vertical scale may be adopted.

2. Field Work

Based on the center line selected in the design office, the final survey is carried out to lay out the final center line of the road in the field.

During this survey necessary additional data for the working drawings are made.

If necessary minor modifications in the final location may be made.

Two operations are involved in this survey, viz:

- Staking out the final center line of the road by means of a continuous transit surveys.
- Detailed levelling.

3. Final Centre-line marking

The decided central line is transferred on the ground by means of a continuous transit survey. As the survey proceeds, staking or pegging of the center line is done. All angles should be measured with a transit. All reference points should be securely located so as not to be disturbed during construction. These reference points are to be transferred on the final plan drawing.

Pegs should be fixed at 50m intervals in plain and rolling terrain and 10 to 20m intervals in hilly terrain. Distances measured along the final line should be continuous following the horizontal curves. At rail and road crossings, the angles subtended by the roads at the center line are measured with the help of a transit. Temporary bench marks are established at suitable intervals (about 300 m) to have vertical controls along the center line.

4. Detailed Survey

Levelling works play an important role as the vertical alignment, earth work computations, drainage design details are to be worked out.

Along the final center line, levels should be taken at all staked stations and at all locations where there are undulations in the ground cross sections.

The following detailed surveys are carried out:

- Cross section and profile.
- Surface drainage area.
- Soil suitability analysis.
- Coarse aggregates availability.
- Right of way investigation.
- Investigation on crossings.

Modern Methods of Surveying

Photogrammetry, often called "Photographic remote sensing" is the science or art of obtaining measurements, needed for highway surveys, by means of photography. It is intended to encompass procedures for photo interpretation and for converting single photographic into composite ones called orthophotographs or mosaics and into maps.

It is based on aerial photographs. It is a basic working tool for the highway engineer. Application of remote sensing is not only helpful in location survey but also in planning, geometric design, rights of way, traffic studies, drainage, soil classification and identification, earthwork measurement, materials location and pavement condition surveys.

Vertical aerial photographs taken with the camera pointed nearly straight down are the most useful for highway mapping purposes. The area to be covered is photographed in parallel with the individual pictures lapped both in the direction of flight (end-lap) and between successive runs (side-lap).

For stereoscopic uses, end-lap must be greater than half the picture width in order that the center of one photograph is included in both adjacent photographs. Side-lap should be around 25% and percentages less than 15% more than 35% are unacceptable. Overlap can be stated in terms of acceptable base to height and width to height ratios. Height from which photographs is to be taken depend on the maps to be made.

The factors to be considered for the purpose of preparing maps are:

- Local length of aerial camera.
- Desired combination of map scale.
- Contour intervals.
- Ratio of map scale to photograph scale.

Several instruments are available for converting data from aerial photographs to maps. They are multiplex, stereoscopic plotters and autograph.

It is possible with any of these instruments to produce an accurate map showing all natural and artificial features. Also, contours may be drawn or spot elevations are determined.

2

Geometric Design of Highways

2.1 Typical Cross Sections of Urban and Rural Roads

The width of a roadway is an important design consideration to ensure that it is appropriately sized to serve its function.

Two major roadway categories established are as follows:

- Rural Road Standards.
- Urban Road Standards.

Urban Road Standards will serve those areas which tend to be more developed and need to provide for multiple users (bicyclists, pedestrians, parallel parking, etc.). Rural roads will primarily serve only vehicular traffic. Cross sections are created to provide a visual guide depicting the initial, interim and ultimate phase cross sections for these road classifications. Threshold daily traffic can be used as a guide for determining which cross section is most applicable.

1. Rural Road Standards

The rural roadways will not typically require curb and gutter or sidewalk. Widths of lanes and shoulders will vary depending upon the specific classification and the potential traffic volume which the roadway may carry.

Roads carrying fewer than 200 vehicles per day need not be paved or treated for dust control. Need for paved shoulders are dependent upon the level of traffic and safety. Final design and construction details will be determined by the Public Works Department.

Final Design and construction criteria taken into consideration may include use of the roadway, density of development, topographical characteristics and nearby development. For construction in which only a portion of the ultimate cross section is intended to be completed, the partial design will need to allow for the eventual widening to the ultimate cross section.

The design for the partial or interim cross section roadway will incorporate ultimate design information to ensure that the first phase of roadway construction is appropriate and would not need to be removed at a future date when the full width cross section is completed.

2. Urban Road Standards

Three roadway classifications are identified for those areas that are associated with the community's urban growth areas. They include arterial, collector and local street classifications.

Urban road standards will include 12-foot lanes, sidewalk and curb & gutter arterials and collectors will also include a striped bike lane. Turn lanes may be necessary as determined by the County. As almost all of the municipalities have different right-of-way cross sections adopted for their community, it makes it very difficult for the County to match them all.

The philosophy was to encourage a baseline amount of right-of-way reservation that could ensure the adjacent community enough area for coordination of future roadway improvements.

2.2 Cross Sectional Elements

Highway Cross Section Elements

Developing a roadway design focuses on the selection and configuration of the elements that comprise the roadway cross section. Once defined these elements are collectively referred to as the typical sections. Cross section elements directly relate to the number of lanes to be provided, the width of pavement, location of shoulders or curbs, medians, slopes, cambers, embankment, slope, drains, etc. Each of the elements is discussed in the following paragraphs.

Right of Way

Right of way is the width of land acquired permanently to be designated as Road Land. Road land requirement depends on the future development of the highway based on the traffic load.

The factors that have influence in taking a decision are:

- Width of formation.
- Shoulders.
- Class of road.
- Expected traffic.
- Availability of land.
- Availability of funds.

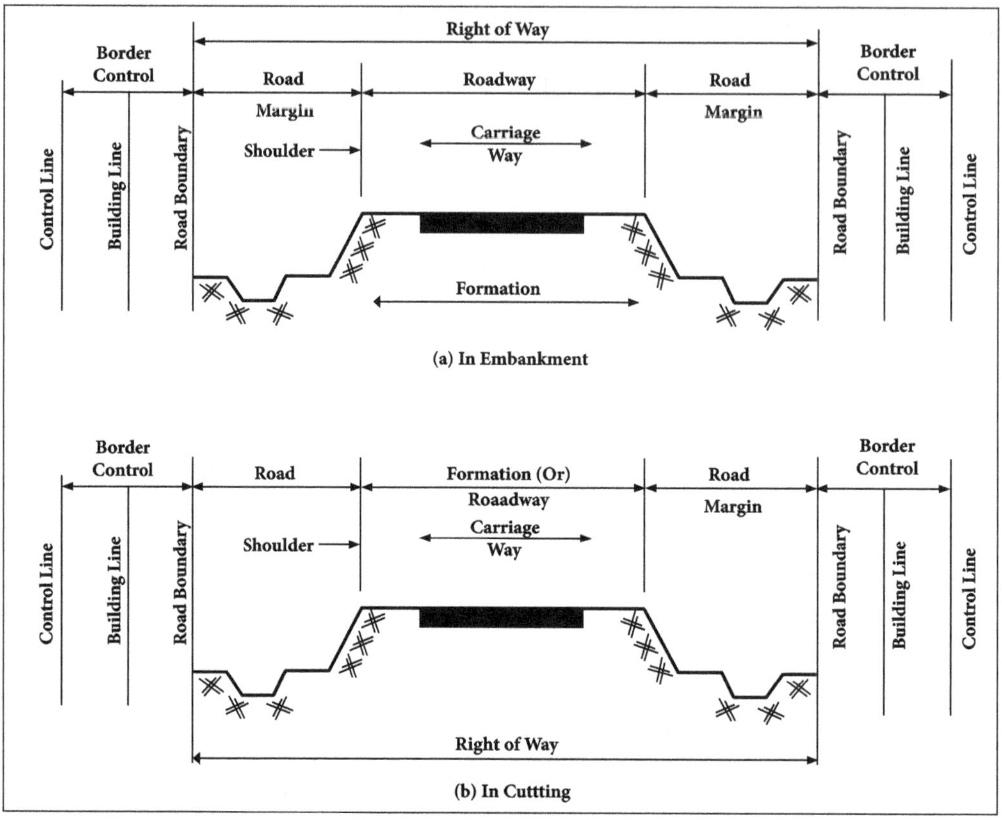

Right of way: (a) In Embankment, (b) In Cutting.

This additional strip of land is required to dump the surplus excavated earth as spoil bank or for obtaining extra earth for road work. Figure shows the position of right of way in embankment and in cutting. Table shows the requirement of right of way for different types of highways on various terrains as per IRC specifications.

Table: Requirement of Right of Way

Road Classification	Plain and Rolling Terrain m				Mountain and Steep Terrain m	
	Open areas		Built - up areas		Open areas	Built - up areas
	Normal	Range	Normal	Range	Normal	Normal
National and State Highways	45	30-60	30	30-60	24	20
Major District Roads	25	25-30	20	15-25	18	15
Other District Roads	15	15-25	15	15-20	15	12
Village Roads	12	12-18	10	10-15	9	9

Border Controls

In order to reserve sufficient space for future improvement of roads, the building construction activities have to be controlled on either side. Building activities should not be allowed beyond building line.

Table: Recommended Border Controls.

Road Classification	Plain and Rolling Terrain m			Mountain and Steep Terrain m	
	Open areas		Built - up areas	Distance between building line and road boundary in m	
	Overall width between building lines, m	Overall width between control lines, m	Distance between building line and road boundary, in m	Open areas	Built-up areas
National and State Highways	80	150	3 to 6	3 to 5	3 to 5
Major District Roads	50	100	3 to 5	3 to 5	3 to 5
Other District Roads	25/30	35	3 to 5	3 to 5	3 to 5
Village Roads	25	30	3 to 5	3 to 5	3 to 5

If the land width is equal to the width between building lines indicated in this column, the building lines should be set-back 2.5 m from the road land building.

The overall requirement of width as per IRC standards are presented in Table. The normal land width required for the National and State Highways on open plain terrain is 45 m and the maximum land width is 60 m. The corresponding width between lines and control lines are respectively 80 m and 150 m. The setback allowed are 10 m and 45 m beyond the road boundary lines keeping the maximum road width.

Carriageway

The pavement width or carriage width depends on the width of traffic lane and number of lanes. Carriage way accommodating one line of traffic movement is called as a traffic lane. Width of the lane is decided based on the width of vehicle and the minimum side clearance which is provided for safety considerations.

Increase of side clearance may allow more vehicles with high speed. Keeping these aspects in view a carriage way width of 3.8 m is considered desirable for a single lane for vehicle of maximum width 2.44 m.

For two or more lanes, the carriage width may be 3.5 m per lane. As per IRC specifications the maximum width of vehicle is 2.44 m. For a single lane, the carriage width is 3.8 m and the side clearance is 0.68 m. For a two lane 7.0 m is provided as carriage way and 0.53 m as side clearance and 1.06 m in between vehicles. The width of carriage way for different lanes or roads is given in Table as per IRC standards.

Table: Recommended Width of Carriageway.

Class of Road	Width of Carriageway
Single lane	3.75 m for all roads; may be decreased to 3 m on village roads
Two lanes without raised kerbs	7.0 m
Two lanes with raised kerbs	7.5 m
Intermediate carriage way (except on important roads)	5.5 m

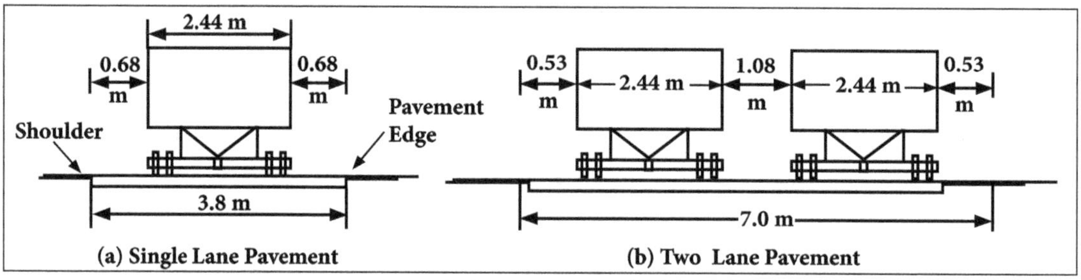

Lateral Placement of Vehicles.

In some highways traffic separators or meridian are provided. In such cases the road width depends on the pavement width (or the lane width and number of lanes) and the width of traffic separators. The width of pavement is increased on horizontal curves.

Kerbs

Kerbs indicate the boundary between the pavement and shoulders. Kerbs are provided on islands or footpaths. Based on their functions, kerbs may be divided into three groups as given below:

1. Class I or Low or Mountable Kerb

It helps the driver to enter the shoulder area with little difficulty. The height of this class of kerbs is 7 to 8 cm. kerbs are also useful for longitudinal drainage system.

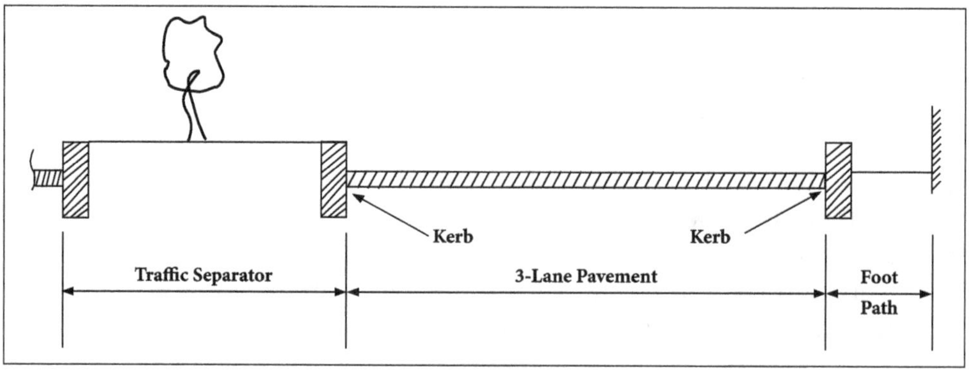

Kerb and Traffic Separator.

2. Class II or Low Speed Barrier or Urban Parking Kerb

It consists of a vertical face of 15 to 20 cm high with 2.5 cm batter. This shape of arrangement is made in the kerb to prevent scraping of tyres.

This type prevents encroachment of slow speed vehicles or parking vehicles to the foot path or parking space. In emergency conditions, the driver can drive over through this curb.

3. Class III or High Speed Barrier

It is considerably high with height of 23 to 45 cm or more so as to prevent vehicles leaving the roadway under any circumstances. It is particularly used in critical locations such as bridges or hill roads. Kerbs are provided to offer lateral stability to granular base courses and flexible pavements. Submerged kerbs are sometimes provided at pavement edges between the pavement edge and shoulders.

Shoulders

Road margin is the term collectively used for the facilities provided on either side of a road pavement, viz., shoulders, parking lane, frontage road, driveway, cycle track, foot path, guard rail and embankment slopes.

Shoulder is that portion of the roadway which lies between the outer edge of the carriage way and the inner edge of the kerb, ditch gutter or slope.

Shoulders are made out of earth, which are of very narrow width of 1.2 m. It can be used only in fair weather. Width recommended by IRC is 2.5 m. larger shoulder widths are now being recommended.

Shoulders are basically provided to serve as an emergency lane for vehicles compelled to be taken out of the pavement and roadway and also serve for vehicles which have broken down.

Factors to be considered for shoulders:

- It should have adequate bearing capacity so as to support loaded trucks even in wet weather.
- Its use increases rapidly with a decrease in pavement width below 7 m.
- Well maintained grass shoulders have the same effect as that of gravel shoulders with respect to the effect on transverse position.
- Surface of shoulders should be rougher than the traffic lane so as to discourage the drivers.
- To distinctly identify, the colour of the shoulder should be different from that of the pavement.

- Bituminous treated shoulders with width more than 1.2 m adjacent to a 7.0 m pavement serve to increase the effective width by 0.6 m.
- The width of shoulder should be around one-half the difference between the roadway width and carriage way width.

Foot Paths

Foot paths also called as sidewalks are provided in roads constructed in urban areas when the vehicular and pedestrian traffic are heavy. It is provided to give adequate protection to pedestrians and thereby decrease accidents. Sidewalks are provided on either side of a road and the minimum width is 1.3 m. Footpaths should be very attractive such that the pedestrians will prefer to use the foot path than the roads.

Separators

Separators are provided to prevent the head on collision between two vehicles moving in opposite directions in the adjacent lanes.

Separators may be in the form of Pavement markings, Physical dividers or area separators.

Out of these three, pavement marking is the cheapest. Physical separators like mechanical separators should be designed in such a way that the vehicles should not be damaged. Wide area separators of 8 to 14 m width may be considered depending on the availability of land.

To reduce the head light glare, a minimum width of 6 m is required. Figure shows kerb and traffic separator. Indian Roads Congress recommends a minimum desirable width of 5.0 m for rural highways.

If adequate land is not available this may be restricted to 3.0 m. On long bridges the separators are reduced to 1.2 to 1.5 m. The separators should be of uniform width on a particular road.

Camber

Also called as cross-fall, is the convexity provided to the cross section of the surface of carriage way. It is the difference in level between the highest point, known as the crown usually located at the center of the carriage and the edge.

Camber is provided so as to perform the following activities:

- To drain surface water.
- To separate the traffic in two opposite directions.
- To improve the appearance of the road.

First reason to provide a camber is very much important and this will require a profile which can remove water quickly to the gutters. Amount of camber to be provided depends on the intensity of rainfalls, rate of run-off and type of road structure.

Second reason provides a natural separating point for the traffic. Third reason adds to the better appearance of cross section. A steep camber is to be preferred in areas of heavy rainfall and for a poor surface material used for the road.

For a better surface material used for the road instruction, such as a concrete road, a flat camber may be enough. Table below shows the minimum camber requirement as per Indian Roads Congress for different types of highway surfaces.

Table: Camber/Cross-fall for different road surfaces.

S. No	Surface Type	Camber/Cross-fall
1.	Cement concrete or High type bituminous surface	1.7 to 2.0 % (1 in 60 to 1 in 50)
2.	Bituminous surface	2.0 to 2.5 % (1 in 50 to 1 in 40)
3.	Water Bound Macadam, Gravel surface	2.5 to 3.0 % (1 in 40 to 1 in 33)
4.	Earth surface	3.0 to 4.0 % (1 in 33 to 1 in 25)

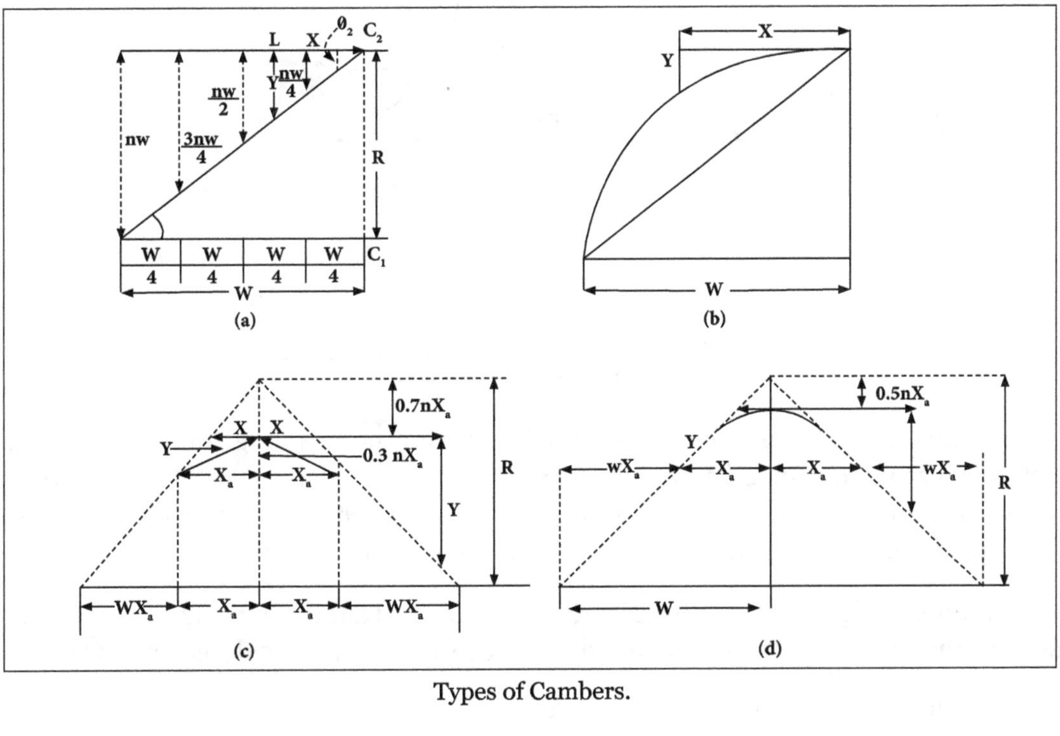

Types of Cambers.

2.3 Sight Distances

Visibility is an important requirement for the safety of travel on highways. For fast moving vehicles and travel during night times, the visibility is of paramount importance.

Visible road length should be sufficient enough such that the driver of a vehicle will be in a position to see any obstruction on the road without any difficulty. There should be adequate length of road visible ahead to the driver at any distance. This length of road is called as sight distance.

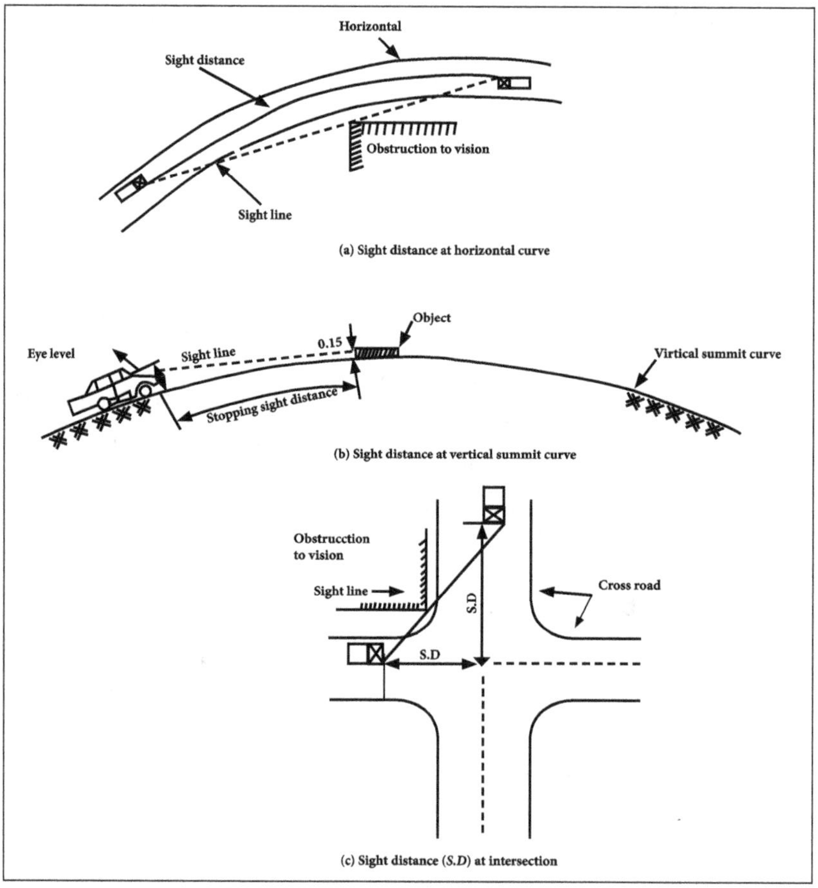

Sight Distance Restrictions.

Sight distance is the actual distance available from a point along the road surface at which a driver from a specified height from the road has visibility of a moving or stationary object on the road. Examples of sight distance restriction on horizontal curves or on vertical curves or at road intersections are shown in figure.

Requirements to sight distance to drivers is needed for traffic control and for geometric design. In the geometric design, three sight distance situations considered are given below.

- Stopping or absolute minimum sight distance.

- Safe overtaking or passing sight distance.
- Safe sight distance at uncontrolled intersection.

Stopping Sight Distance

A driver on a vehicle driving at the design speed on a highway should have sufficient sight distance or length of road visible ahead to stop the vehicle in case of any obstruction on the road without collision. The stopping sight distance is the distance covered by a vehicle from the instance a danger is fully realized by the driver to the actual stop. This is also called as non-passing sight distance.

The sight distance available on a road to a driver depends on the following factors:

- Features of the road ahead.
- Height of the driver's eye above the road surface.
- Height of the object above the road surface.

Stopping of a motor vehicle on a road after seeing an obstruction depends on the following factors:

- Total reaction time of the driver.
- Speed of vehicle.
- Brake efficiency of the vehicle.
- Frictional resistance between road and the tyres.
- Gradient of the road.

1. Total Reaction Time

It is the time taken from the time the object is seen by the driver to the time the brakes are effectively applied.

As per the PIEV theory the total reaction time comprises of four parts, viz.,

- Perception,
- Intellection,
- Emotion,
- Volition.

Perception time is the time required for transmitting the sensation received by the driver's eyes or ears to the brain through the nervous system and spinal cord.

Intellection time is the time required for realizing the situation. It is also the time taken for comparisons of different thoughts and alterations, regrouping and registering new sensations.

Emotion time is the time elapsed during emotional sensations and disturbance such as fear, anger or any other emotional feelings with reference to the situation. Volition time is the time taken for the final action. The PIEV process is illustrated in figure.

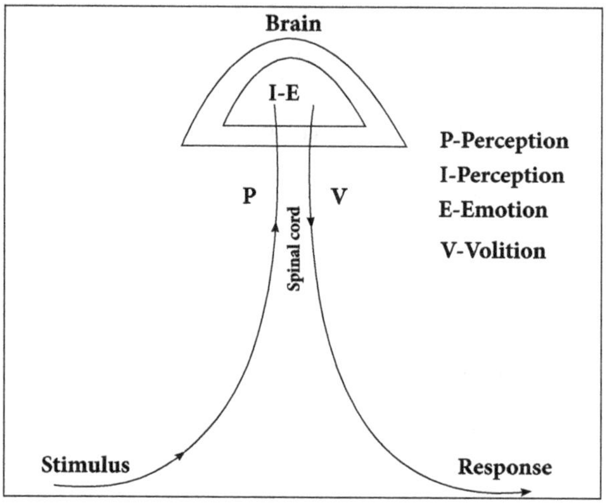

Reaction time and PIEV process.

The PIEV time of a driver depends on several factors such as mood, temperament and physical conditions of the driver, environmental factors, etc. For an average driver, the total reaction time varies from 0.5 seconds to 4 seconds depending on the complexity of the problem.

2. Speed of Vehicle

The distance moved by the vehicle during the reaction time is a function of the speed of the vehicle. After the application of the brake the distance moved by the vehicle before coming to a stop is based on the initial speed of the vehicle. Higher the speed more will be the stopping distance.

3. Brake Efficiency

If the brakes are in good condition, the wheels are fully locked once the brake is applied preventing the wheels from rotating. It will lead to 100 %skidding which is not generally desirable and dangerous. In order to avoid skidding, the force due to brakes should not exceed the frictional force between the wheels and tyre.

4. Frictional Resistance between Road and Tyre

The coefficient of friction between the road surface and the tyres of a vehicle depends on the following:

- The type and the condition of road surface.

- Presence of moisture, mud, snow, etc. on the road surface.
- Conditions of treads of tyres.
- Tyre pressure.

Indian Roads Congress recommends a value of 0.40 at 20 km/hr to 0.35 at 100 km/hr for the frictional coefficient. Equating the work done by friction to the difference in the kinetic energies of the vehicle at the time of applying brakes and at the time of stopping, gives the braking distance. In general a rate of retardation of 3.93 m/sec^2 is considered comfortable to passengers.

5. Gradient of the Road

Depending on ascending or descending gradient, the gravity adds to or against braking action.

In ascending gradient, the gravity adds to the braking action and the breaking distance is decreased.

Components of Stopping Sight Distance

The total stopping distance of a vehicle consists of the following components:

- Lag distance.
- Braking distance.

1. Lag Distance

The lag or reaction distance is the distance moved by a vehicle during the total reaction time.

2. Braking Distance

Braking distance may be computed by equating the work done in stopping the vehicle and the kinetic energy for a level road. Several factors contribute for the friction on the road. Let f be the coefficient of friction. The friction decreases with the increase in speed of the vehicle.

IRC has recommended values of friction for different design speeds (Table).

Table: Coefficient of Friction for Different Speeds.

Speed, km/hr	20 to 30	40	50	60	65	80	100
Longitudinal Coefficient of Friction, f	0.40	0.38	0.37	0.36	0.36	0.35	0.35

Table: SSD values for Different Speeds.

Design Speed, kmph	20	25	30	40	50	60	65	80	100
Safe SSD for Design, m	20	25	30	45	60	80	90	120	180

Problem

1.Let us consider the Design speed = 80 kmph, Total reaction time = 3 sec, Coefficient of friction = 0.6. With the following specifications. Calculate SSD for (i) A two lane road with two - way traffic and (ii) A single lane road with two - way traffic.

Solution:

Given:

Design speed = 80 kmph,

Total reaction time, t = 3 sec,

Coefficient of friction = 0.6.

Formula used:

$$\text{Stopping Distance} = vt + \frac{v^2}{2gf}$$

Assume the road is a level road.

Then Stopping Distance = (Lag distance) + (Braking distance)

$$V = 80\,\text{kmph or } v = \frac{80}{3.6} = 22.22 \text{ m/sec}$$

t= 3 sec, g=9.8 g/cm² and f=0.6

Therefore, $$\text{Stopping Distance} = vt + \frac{v^2}{2gf}$$

$$= 22.22 \times 3 + \frac{(22.22)^2}{2 \times 9.8 \times 0.6}$$

=66.66 + 41098

=108.64 m

i) Stopping Sight Distance for two lane road with two way traffic

= Stopping Distance

= 108.64 m

ii) Stopping Sight Distance for single lane road with two way traffic = 2 × 108.64

= 217.28 m

Overtaking Sight Distance

Overtaking sight distance is the minimum sight distance that should be available for a driver to overtake another vehicle safely on a two-way road.

It is also referred to as safe passing sight distance. Considering the driver's eye position as 1.2 m above road surface, the overtaking sight distance can be defined as the distance along the center line of the road with driver's eye level at a height of 1.2 m above the road.

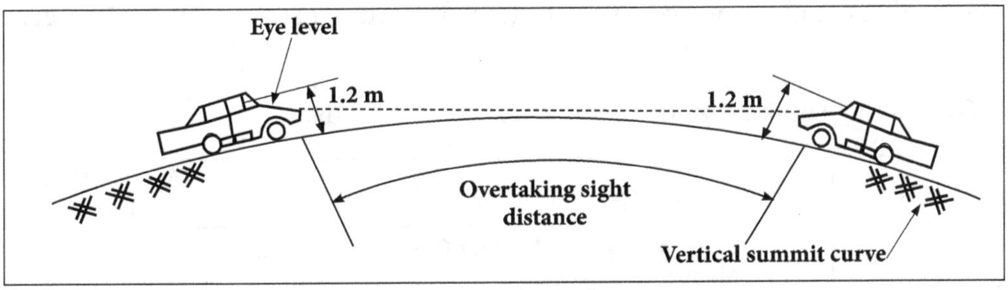

Measurement of overtaking sight distance.

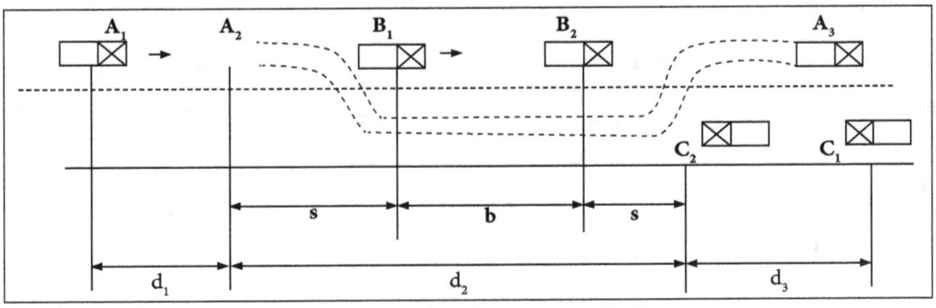

Manoeuvre of overtaking vehicle.

Minimum overtaking sight distance is governed by the following factors:

- Speeds of over-taken and overtaking vehicles.
- Other vehicles coming from the opposite direction.
- Space between overtaken and overtaking vehicles.
- Skill and reaction time of the driver.
- Rate of acceleration of overtaking vehicle.
- Slope of the road.

Table: Maximum overtaking acceleration at different design speeds.

Design Speed		Maximum Overtaking Acceleration	
V kmph	v m/sec	A_a kmph/sec	a m/sec^2
25	6.93	5.0	1.41
30	8.34	4.8	1.30
40	11.10	4.45	1.24
50	13.86	4.00	1.11
65	18.00	3.28	0.92
80	22.20	2.56	0.72
100	27.80	1.92	0.53

The IRC has specified safe overtaking sight distances for various design speeds.

The OSD values thus obtained are rounded off and presented by IRC as given in table.

Table: Overtaking sight distance for various design speeds.

Design Speed kmph	Time Components, Seconds			Safe Overtaking Sight Distance m
	For Overtaking Manoeure	For On-coming Vehicle	Total	
40	9	6	15	165
50	10	7	17	235
60	10.8	7.2	18	300
65	11.5	7.5	19	340
80	12.5	8.5	21	470
100	14	9.0	23	640

Overtaking Zones

It is always desirable to construct roads such that the visibility is available well ahead at every point of the road. It is not feasible in practice. In areas where visibility is not adequate a prohibitive board such as "No passing" or "Overtaking prohibited" may be displayed.

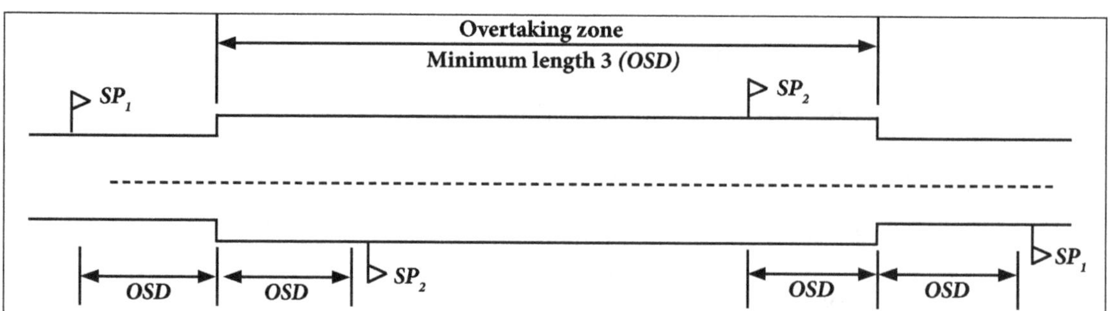

Overtaking Zone.

SP_1 - Sign Post - Overtaking zone ahead.

SP_2 - Sign Post - End of overtaking zone.

Vehicles moving at design speed should be given opportunity to overtake along the road at frequent intervals. Areas or zones which are meant for overtaking are called, overtaking zones.

The minimum length of overtaking zone should be three times the safe over-taking distance, i.e., $3(d_1 + d_2)$ for one-way roads and $3(d_1+d_2+d_3)$ for two-way roads. It is desirable that the length of overtaking zones is kept five times the overtaking sight distance. Adequate sign posts should be installed on either side of the overtaking zone. The sign post should be kept at a distance of (d_1+d_2) for one-way roads and $(d_1+d_2+d_3)$ for two-way roads.

Problem

1.Let us consider, If the acceleration of overtaking vehicle is 0.98 m/sec² and speed of overtaking and overtaken vehicles are 80 and 50 kmph respectively, on a two-way traffic road then with these assumptions calculate,

- Safe overtaking sight distance.
- Minimum length of overtaking zone and show the positions of the sign.

Solution:

Given:

Acceleration of overtaking vehicle is 0.98 m/sec^2.

Speed of overtaking and overtaken vehicles is 80 and 50 kmph.

Consider the speed of the overtaking vehicle as the design speed, i.e.

$$v = 80\,\text{kmph} = \frac{80}{3.6} = 22.22\,\text{m/sec.}$$

Speed of overtaken vehicle,

$$v_b = 50\,\text{kmph} = \frac{50}{3.6} = 13.89\,\text{m/sec}$$

Acceleration, a = 0.98 m/sec^2.

$$OSD = d_1 + d_2 + d_3$$

Assume reaction time as 2 sec., i.e., t = 2 sec.

Therefore $d_1 = vbt$

$= 13.89 \times 2 = 27.78$ m.

$s = (0.7\ vb + 6)$

$= 0.7 \times 13.89 + 6 = 13.723$ m

Time, $T = \sqrt{\frac{4 \times 13.723}{0.98}} = 7.6 \sec$

$d_2 = vbT + 2\ s$

$= 13.89 \times 7.6 + 2 \times 13.723$

$= 127.01$ m

$d_3 = vT$

$= 22.22 \times 7.6$

$= 161.87$ m.

OSD = 27.78 + 127.01 +161.87

= 316.66 m

- i.e., OSD = 317 m.
- Minimum length of OSD (for two-way traffic).

= 3 (OSD)

= 3 × 317 = 951 m

Position of sign posts are shown in figure:

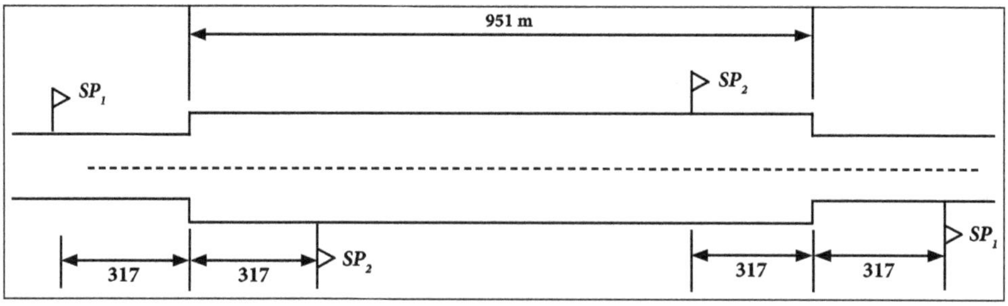

Overtaking Zones.

Sight Distance at Intersections

Uncontrolled intersection is one where there is no automatic signal or any policeman available to control the traffic. Vehicle entering an uncontrolled intersection should have sufficient visibility so as to enable the driver to control his vehicle and to avoid collision with other vehicles moving in from other roads. It is essential that sufficient sight distance is available along the intersecting roads. Uncontrolled intersections may be grouped under two categories:

- Uncontrolled intersections where there are intersecting roads of more or less equal importance.
- Priority intersections consist of a main road intersecting a minor road.

1. Uncontrolled Intersectiont

In these intersections it is essential that there is a clear view across the corners from a sufficient distance so as to avoid collision of vehicles. Existence of some structures or some objects at the corners may obstruct the view. The unobstructed area formed by the lines of vision and sight line, called the sight triangle (ABC), is shown in figure:

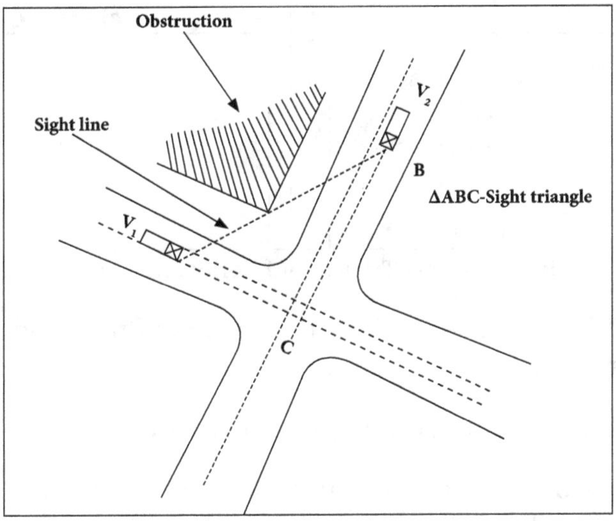

Sight Distance at Uncontrolled Intersection.

Sight distance requirements at uncontrolled intersections may be decided on the basis of three possible conditions:

- Vehicle to Change Speed.
- Vehicle to Stop.
- Vehicle to Cross a Road.

2. Priority Intersection

Here the visibility provided should be such that the vehicles moving towards the main road are able to see the vehicles on the main or major road. The time available should be enough to judge whether the required space is available to cross the main road without any interference of the traffic on the main road. For this purpose, IRC has recommended a minimum visibility distance of 15 m along the minor road.

A driver on a minor road takes about 8 seconds to perceive the traffic conditions on the intersection, to estimate the gap in the vehicle stream, to take a decision about the crossing and to accelerate the vehicle to complete the maneuver. On this basis, the sight triangle at priority intersections should be formed with BC measuring 15 m along the minor road.

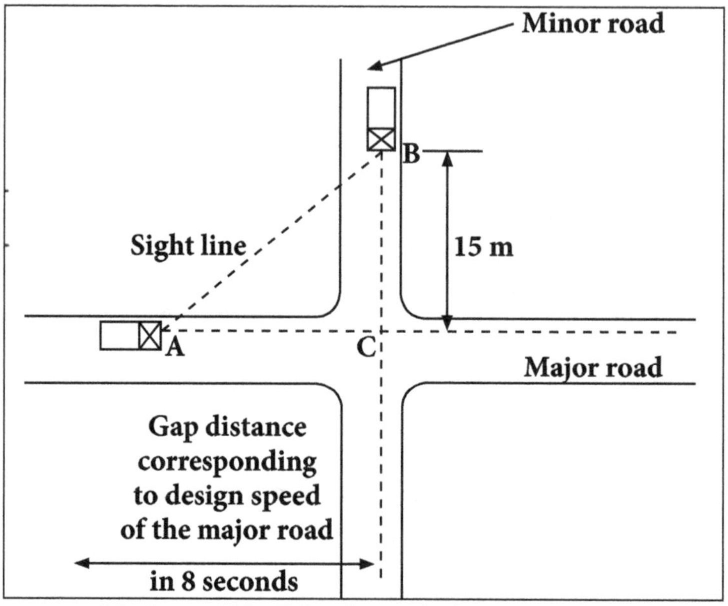

Minimum Sight Triangle at Priority Intersection.

Now the distance AC corresponds to 8 seconds reaction time and the design speed of vehicle on the major road.

Intermediate and Head Light Sight Distance

Apart from the three stopping sight distances discussed so far, the following sight distances are recommended by IRC for consideration in highway design.

- Intermediate Sight Distance:
    - Intermediate Sight Distance is taken as twice the Stopping Sight Distance.
- Head Light Sight Distance:
    - This is the visible distance for a driver during night time under the illumination of head lights.

2.4 Horizontal Curves

Horizontal Curves

It is provided in the horizontal plane connecting two straight alignments.

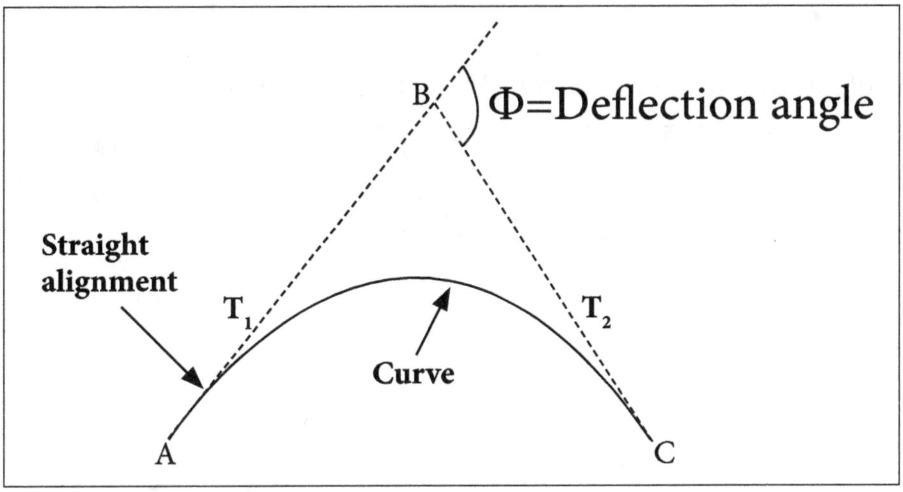

Horizontal Curve.

Types of Horizontal Curves

Horizontal curves may be classified as:

- Simple curve.
- Compound curve.
- Reverse curve.
- Transition curve.

Simple Curve

It consists of a single arc with a constant radius connecting two straights or tangents.

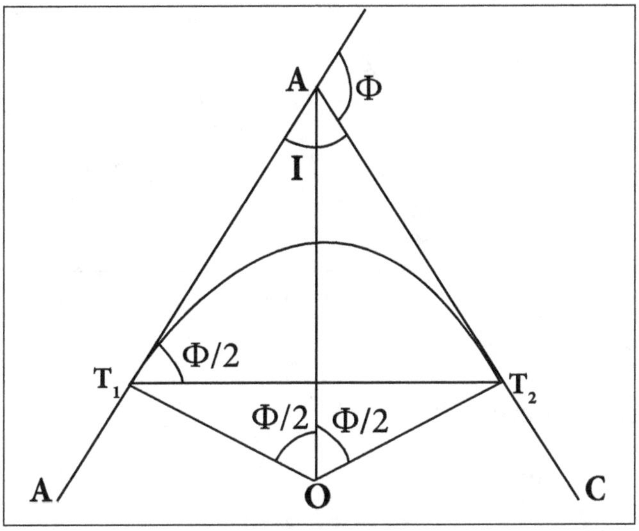

I = Interior angle
Simple Circular Curve.

Compound Curve

It consists of more than one radius connecting two intersecting straights.

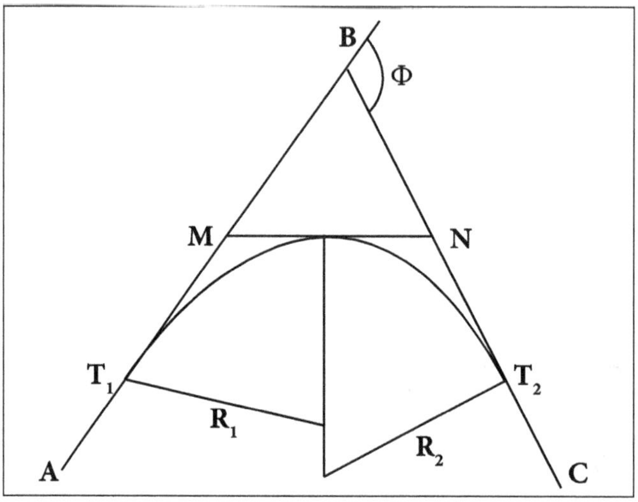

R_1 - Radius of first circular arc.
R_2 - Radius of second circular arc.

Reverse Curve

When two curves of different or equal radius are bending in opposite directions then it is called a reverse curve. Reverse curves have one common tangent.

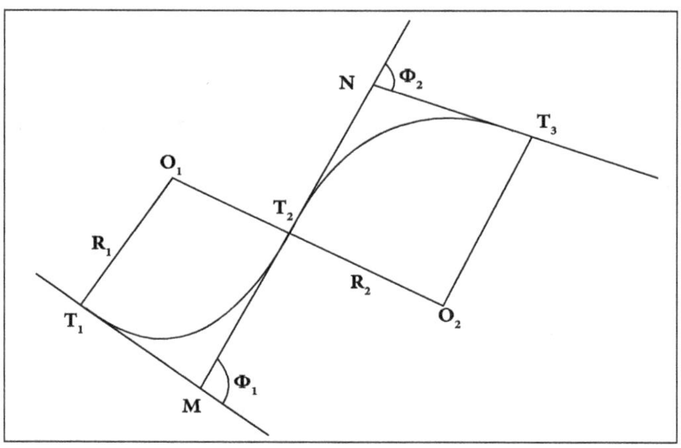

Reverse Curve.

Φ_1, Φ_2 - Deflection angle.

R_1 - Radius of first circular arc.

R_2 - Radius of second circular arc.

MN - Common tangent.

Transition Curve

It is a curve having a gradual varying radius. Example: a curve with infinite radius in the beginning and varying gradually to a finite radius.

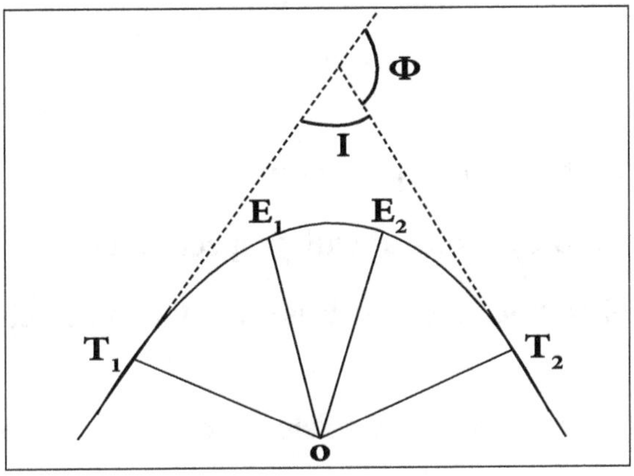

Transition Curve.

E_1E_2 - Circular curve.

T_1E_1, T_2E_2 - Transition curves.

I = Interior angle.

Φ - Deflection angle.

Super Elevation

To counter balance the centrifugal force, the outer edge of the road is raised which is known as the super elevation or cant or banking. Transverse slope is provided throughout the length of the horizontal curve.

It is expressed as the ratio of the height of the outer edge with respect to the horizontal width. Figure shows the details of super elevation:

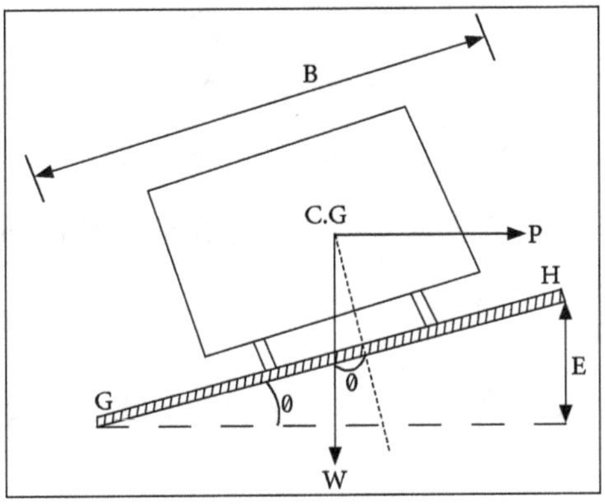

Super Elevated Section of Pavement.

$$e = \frac{HI}{GI} = \tan\theta.$$

The value of e is very small and of the order of 0.07, hence tan θ = sin θ, then,

$$e = \tan\theta = \sin\theta \frac{HI}{GH} = \frac{E}{B}$$

Which is the height to width ratio of the pavement.

Total rise in outer edge of the pavement with respect to the inner edge HI=E = e B.

Where, e is the super elevation rate and E is the total super elevated height of outer edge.

Now all the forces acting are summarized below along with the components.

- The centrifugal force, $P = \frac{Wv^2}{gR}$, acting horizontally through the C.G.
- The weight W of the vehicle acting vertically downwards through the C.G.
- The friction developed between the wheels and the pavement, FA and FB acting parallel to pavement elevated position:

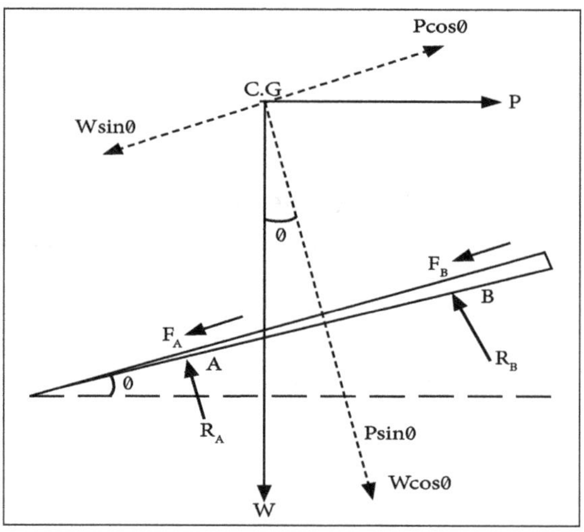

Forces acting on a moving vehicle on a super elevated pavement.

Considering the equilibrium of forces acting parallel to the pavement,

$$P\cos\theta = W\sin\theta + Fa + Fb$$

Considering the equilibrium of forces acting perpendicular to the pavement,

$$RA + RB = W\cos\theta + P\sin\theta$$

Rewriting the above Equation,

$$P\cos\theta = W\sin\theta + f(RA + RB)$$

Substituting for RA + RB from above equation we have,

$$P\cos\theta = W\sin\theta + f(W\cos\theta + P\sin\theta)$$

$$\text{i.e., } P(\cos\theta - f\sin\theta) = W\sin\theta + f\,W\cos\theta$$

Dividing by W cos θ throughout,

$$\frac{P}{W}(1 - f\tan\theta) = \tan\theta + f$$

$$\text{i.e., } \frac{P}{W} = \frac{\tan\theta + f}{1 - f\tan\theta}$$

The coefficient of lateral friction is generally taken as 0.15 and tan o seldom exceeds 1/15. Hence f tan θ is less than 0.01. Thus the denominator of the centrifugal ratio 1-f tan θ = 0.99 which can be approximated to 1, then,

$$\frac{P}{W} = \tan\theta + f = e + f$$

$$\text{But, } \frac{P}{W} = \frac{v^2}{gR}$$

$$\text{i.e., } e + f = \frac{v^2}{gR}$$

The above equation can be represented taking V in kmph as,

$$e + f = \frac{(0.278V)^2}{9.8R} = \frac{V^2}{127R}$$

Neglecting the coefficient of friction, the equilibrium super elevation required to countract the centrifugal force fully is given as,

$$e = \frac{v^2}{gR}$$

$$e = \frac{V^2}{127R}$$

Once this super elevation is provided, the pressure on the inner and outer wheels will be equal. This condition will result in a very high value of super elevation. As lateral friction cannot be ignored completely f is always taken into account.

In some field situations it may not be possible to provide super elevation (i.e., e = 0), then the later friction alone should counteract the centrifugal force. In such situations:

$$f = \frac{v}{gR}$$

Or,

$$v = \sqrt{fgR}$$

And

$$f = \frac{V^2}{127R}$$

Or,

$$V = \sqrt{127fR}$$

1. Limits of Super Elevation

Maximum

Amount of superelevation increases with increase in speed and with decrease in radius of the curve, for a constant value of f. Maximum e should be fixed on roads with mixed traffic.

Indian Roads Congress has fixed a maximum limit of superelevation as 7 % in plain and rolling terrain and with mixed traffic. A maximum of 10 % is allowed on hill roads.

Minimum

From drainage point of view a certain superelevation has to be provided. If the calculated e is less than or equal to camber of the road surface then e may be limited to camber of the road on the curve.

Table: List of Radii Beyond which Superelevation is not required.

Design Speed (kmph)	Radius (Metres) for Camber of				
	4%	3%	2.5%	2%	1.7%
20	50	60	70	90	100
25	70	90	110	140	150

30	100	130	160	200	240
35	140	180	220	270	320
40	180	240	280	350	420
50	280	370	450	550	650
65	470	620	750	950	1100
80	700	950	1100	1400	1700
100	1100	1500	1800	2200	2600

The IRC has recommended for various speed of vehicles, the radii along with camber for which no super elevation is needed.

Design of Super Elevation

In the design of super elevation the following factors are to be considered:

- For a mixed traffic, the super elevation provided for the speedy vehicle with f = 0 is safer for most of the vehicles. This may be inconvenient for slow moving vehicles.
- For a mixed traffic, providing super elevation based on lateral friction would be unsafe for fast moving vehicles.
- Considering both these aspects, as a compromise, the centrifugal force due to 75 % of the design speed and limiting the maximum superelevation to 0.07 are adopted.

For hilly roads, the maximum superelevation may be taken as 0.10. Following steps are to be adopted in the design of a superelevation for a pavement:

Step 1:

Superelevation for 75 % of design speed (v m/sec or V kmph) is calculated neglecting the friction (i.e., taking f = 0).

$$e = \left(\frac{(0.75v)^2}{gR} \right)$$

Or,

$$e = \frac{(0.75V)^2}{127R} = \frac{V^2}{225R}$$

Step 2:

If the computed value of e is less than 7 %, then computed value is provided.

If the value is greater than 7 %, then provide 7 % superelevation and steps 3 and 4 are followed.

Step 3:

Compute the coefficient of friction developed for the maximum value of e= 0.07 at the full value of design speed from,

$$f = \left(\frac{v^2}{8R} - 0.07\right)$$

Or,

$$f = \left(\frac{v^2}{127R} - 0.07\right)$$

If the computed f value is less than 0.15, then the superelevation provided (7 %), is safe for the design speed. If not, the restricted speed limiting f = 0.15, is computed as explained in Step 4.

Step 4:

The allowable speed for the given curve for the maximum superelevation of 7 % and friction of 0.15, that is:

$$e + f = 0.07 + 0.15 \div 0.22$$

i.e., $$e + f = \frac{v_a^2}{8R}$$

or, $$e + f = \frac{V_a^2}{127R}$$

If the calculated allowable speed is higher than the design speed then the design is adequate and the super elevation, e = 0.07 is provided.

If the calculated allowable speed is less than the design speed, the speed is limited to V_a kmph as calculated above. Speed limit signs are installed at appropriate places.

In the case of important highways, speed restrictions cannot be made. Instead the curve should be re-aligned, if adequate land is available, with a large radius so that the design speed can be maintained.

Problem

1.Let us consider the design speed on a road with a horizontal curve of radius 400 m is 90 km/hr. The coefficient of friction available on the road may be taken as 0.15. Calculate:

- Superelevation when full lateral friction has developed.
- Coefficient of friction when no super elevation is provided.
- Superelevation for equal pressure at inner and outer wheels.

Solution:

Given:

Horizontal curve of radius 400 m is 90 km/hr.

Coefficient of friction available on the road is 0.15.

- We know that,

$$e + f = \frac{V^2}{127R}$$

Substituting the known values,

$$e + 0.15 = \frac{(90)^2}{127 \times 400} = 0.1595$$

e = 0.1595-0.15 = 0.0095.

i.e., Superelevation rate = 0.95%.

- No superelevation is provided, i.e., e = 0.

Then, coefficient of friction, $f = \frac{V^2}{127R} = \frac{90^2}{127 \times 400} = 0.1595$

- When pressure on the inner and outer wheels is equilibrium condition, superelevation for which f = 0.

$$\therefore e = \frac{V^2}{127R} = \frac{(90)^2}{127 \times 400} = 0.1595$$

Superelevation is 15.95 % which is more than the maximum permissible percentage of 7% and hence cannot be provided.

Transition Curve

It is also called as an easement curve. It is an arc introduced between a straight and a circular curve or between two arcs of a compound curve. The radius of a transition curve varies from infinity to a fixed value.

Vehicles passing straight into a circular curve experience jerk due to the effect of centrifugal force. Centrifugal force always acts in a direction perpendicular to the axis of rotation.

To counter act the above effect, the outer edge of the road or the rail is rose which is called the superelevation or cant. Superelevation is provided gradually along the transition curve.

Advantages:

- Introduces superelevation gradually from zero at the tangent point to the value on the circular curve.
- Maintains a constant proportional between superelevation and rate of change of curvature.
- Eliminates discomfort, overturning and side slipping of vehicles.
- Eliminates discomfort to passengers.

Requirements:

A transition curve should satisfy the following requirements:

- The transition should be tangential to the straight curve.
- The curvature of the transition curve should be zero at the origin of the straight curve.
- The exact amount of superelevation should be attained at the junction with the circular curve.
- The curvature of the transition curve should increase at the same rate as that of the superelevation.

Details of a Composite Curve

Transition curves are introduced at both ends of a circular curve, the resulting curve is known as a combined or composite curve:

AB - The rear or back tangent.

BC - The forward tangent.

T_1 - Starting point of the curve or first tangent.

T_2 - Ending point of the curve or second tangent.

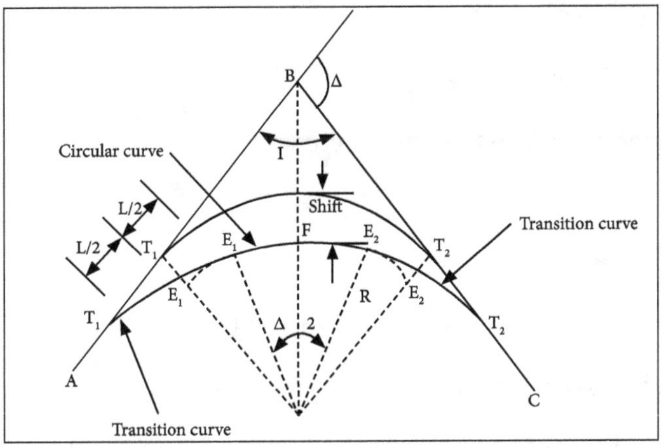

Composite Curve.

D-Deflection angle,

I-Angle of intersection,

ϕ -Spiral angle,

T_1E or T_2D- Length of the transition curve,

EFD-Length of circular curve,

$T'_1E'_1$ or $T'_2E'_1$-Shift of the curve,

BT_1 or BT_2- Total tangent length,

T_1E F DT_2-Length of the combined curve,

Angle $(\Delta - 2\phi)$ - Central angle.

Types of Transition Curves

Euler's Spiral

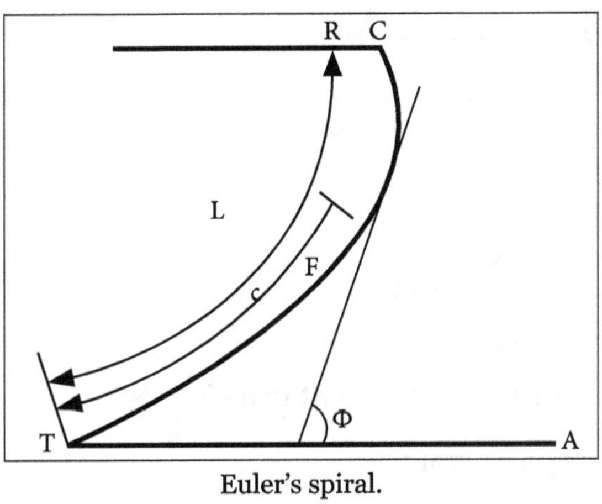

Euler's spiral.

Figure shows an Euler's spiral.

Equation for the above curve, $\phi = \frac{l^2}{2RL}$

This is the ideal equation of the transition curve.

Here,

F - A point on the curve.

ϕ - Angle made by tangent at F with initial tangent TA.

l - Distance between F and T.

L - Total length of transition curve.

R - Radius of circular curve.

Cubical Spiral

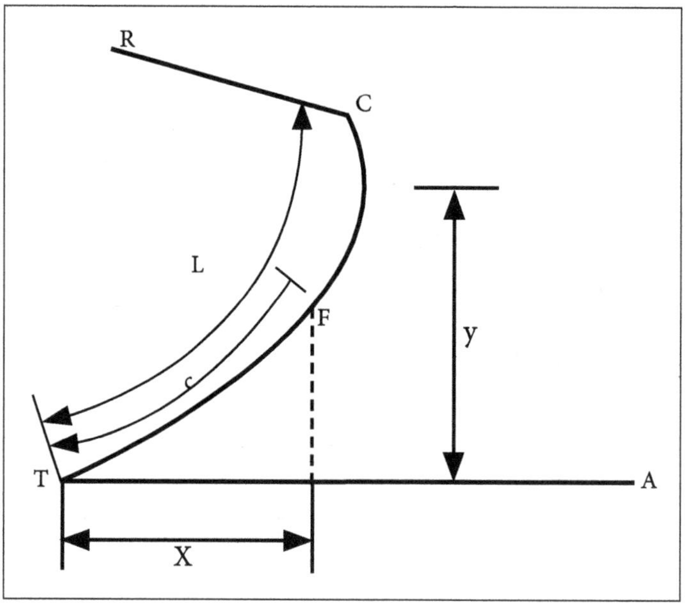

Cubical spiral.

Figure shows a cubical spiral.

The equation of this curve is $y = \frac{l^2}{6RL}$

Where,

y - Perpendicular offset to any point from TA.

l - Distance of the point from T.

L - Total length of transition curve.

R - Radius of circular curve.

Cubic Parabola

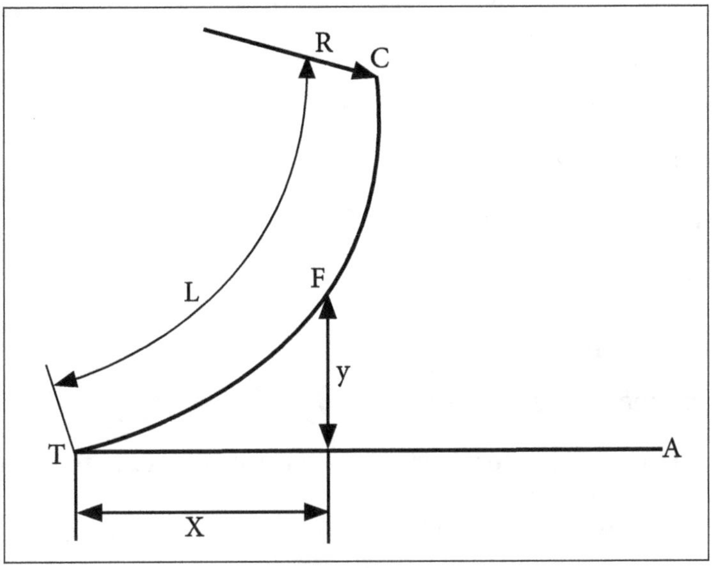

Cubic parabola.

Figure shows a cubic parabola.

$$y = \frac{x^3}{6RL}$$

Where,

x - Horizontal distance between F and T.

y - Perpendicular distance between F and TA.

Cubic parabolas are generally recommended for railways.

Details of Transition Curve

1. Length

Length of a transition curve is calculated based on one of the following considerations:

- By adopting a particular rate of superelevation.
- By considering arbitrary rate of superelevation.
- By considering rate of change of radial acceleration.

By adopting a particular rate of superelevation.

The definite rate of superelevation varies from 1 in 300 to 1 in 1000.

Length of transition curve, $L = \frac{ne}{100} m$.

Where,

e - Amount of superelevation in centimeters.

n - Rate of superelevation.

By considering arbitrary rate of superelevation,

Length of transition curve, $L = \frac{e \times V}{x}$.

Where, e - Amount of superelevation.

V - Speed in meters per second.

x - Time rate in centimeters per second (varies from 2.5 to 5 cm/sec).

By considering rate of change of radial acceleration,

Radius acceleration on circular curve = $\frac{V^2}{R}$.

Time taken by the vehicle to cover transition curve = $\frac{L}{V} s$.

If the change of radial acceleration is K m/s^3 then,

Time taken to attain maximum radial acceleration = $\frac{V^2}{KR} s$.

From equation,

$$\frac{L}{V} = \frac{V^2}{KR}$$

$$\text{i.e., } L = \frac{V^3}{KR}$$

2. Spiral Angle

From the intrinsic equation of the ideal transition curve, we have,

$$\phi_1 = \frac{l^2}{2RL} \text{radians}$$

The angle ϕ is known as the spiral angle.

3. Deflection Angle

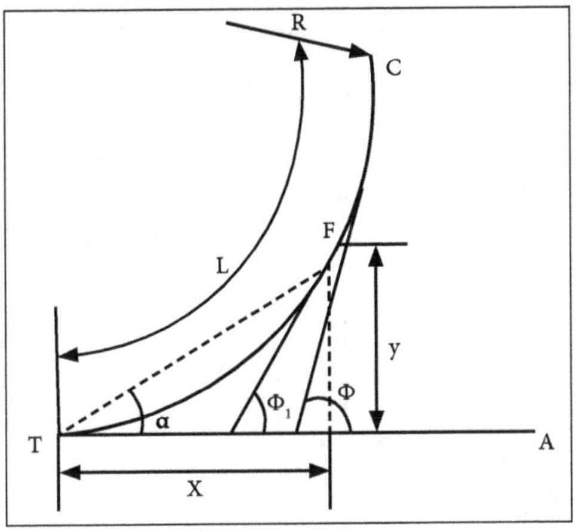

Deflection Angle.

Here,

F - Any point on transition curve.

C - Point of junction of circular and transition curve.

ϕ - Spiral angle.

ϕ - Angle between tangent at F and initial tangent.

α - Deflection angle at F.

x and y - Rectangular coordinates.

R - Radius of circular curve.

i - Length of transition curve between T and F.

Now, $\tan\alpha = \frac{y}{x}$.

From equation for cubical spiral, we know that,

$$y = \frac{l^3}{6RL}(\text{where } x \approx 1)$$

Considering α as small, then $\tan \alpha \approx \alpha$.

$$\therefore \alpha = \frac{1}{x} \times \frac{x^3}{6RL} = \frac{x^2}{6RL} = \frac{l^3}{6RL} \text{radians}$$

$$\alpha = \frac{5731^2}{RL} \text{mins}$$

4. Shift

The distance by which the main curve is moved inwards to accommodate the transition centre is called a shift.

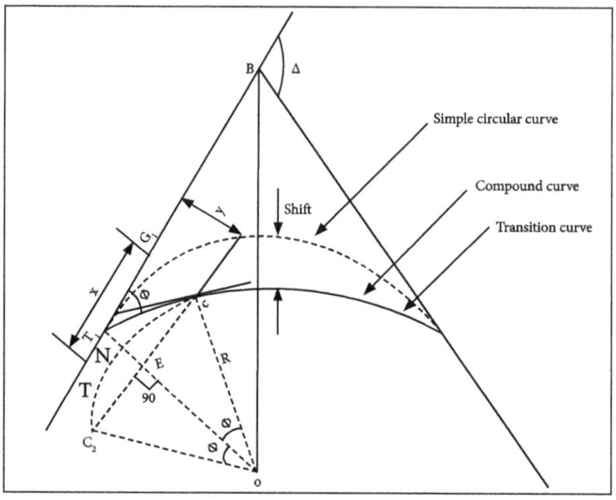

Compound Curve.

Here,

TB - Initial tangent.

T - Tangent point of combined curve.

T_1 - Tangent point of simple circular curve.

S - Shift.

C - Junction point of transition curve with circular curve.

DC - Common tangent.

ϕ - Spiral angle.

N - Intersection point of shift and transition curve.

R - Radius of circular curve.

L - Length of transition curve.

From the intrinsic equation of the ideal transition curve, we have

Spiral angle, $\phi = \dfrac{L}{2R}$

$$CE = R\phi = R \times \frac{L}{2R} = \frac{L}{2}$$

But CN is very nearly equal to CE.

$$CN = \frac{L}{2}$$

$$\text{Hence, } TN = \frac{L}{2}$$

So, the shift T_1E bisects the transition curve at N.

Now, CC_2 is perpendicular to OT_1 and CC_2 is perpendicular to TB.

$$S = C_2T_1 - C_2E$$

(Where, C_2E = versed sine of curve with central angle 2φ).

$$= CC_1 - R(1 - \cos\phi)$$

$$= y - R(1 - \cos\phi)$$

$$\therefore S = y - R \times 2\sin^2\frac{\phi}{2}$$

We know,

$$y = \frac{x^3}{6RL} = \frac{L^3}{6RL} = \frac{L^2}{6R} \quad (x \approx L)$$

Considering ϕ as small, $\sin\phi \approx \phi$.

$$S = \frac{L^2}{6R} - R2\left(\frac{\phi}{2}\right)^2 = \frac{L^2}{6R} - 2R \times \frac{\phi^2}{4}$$

$$= \frac{L^2}{6R} - 2R \times \frac{1}{4}\left(\frac{L}{2R}\right)^2$$

$$= \frac{L^2}{24R}$$

$$\text{i.e., Shift, } S = \frac{L^2}{24R}$$

$$\text{Again, } T_1N = \frac{TN^3}{6RL} = \frac{(L/2)^3}{6RL} = \frac{L_2}{48R}\left(TT_1 = TN = \frac{L}{2}\right)$$

$$\therefore \frac{\text{Shift}}{T_1N} = \frac{L^2N}{24R} \times \frac{48R}{L^2} = 2$$

Hence, $T_1N = \frac{1}{2}$ Shift $= \frac{1}{2} T_1E$

5. Elements of Combined Curve

- The length of the transition curve may be calculated from the formula, if not given.

$$L = \frac{V^3}{KR}$$

- Shift, $S = \frac{L^2}{24R}$.

Where, R = Radius of circular curve.

- Tangent length of the combined curve (cubic parabola as transit curve).

$$BT_1' = BT_1 + T_1, T_1' = OT_1 \tan\frac{\Delta}{2} + \frac{L}{2}$$

$$= (R+S)\tan\frac{\Delta}{2} + \frac{L}{2}$$

Where, Δ = Deflection angle.

- Spiral angle.

$$\phi = \frac{L}{2R} \text{radians} = \frac{L}{2R} \times \frac{180^\circ}{\pi} \text{deg.}$$

- Central angle for circular curve = $\Delta - 2\phi$.
- Length of circular curve = $\frac{\pi R(\Delta - 2\phi)}{180^\circ}$.
- Length of combined curve = $T_1E_1 + E_1E_2 + E_2T_2$.

$$= L + \frac{\pi R(\Delta - 2\phi)}{180^\circ} + L = \frac{\pi R(\Delta - 2\phi)}{180^\circ} + 2L$$

- Chainage of point of start of the curve is given by= T_1 = chainage of B - BT_1.
- Chainage of E_1 = Chainage of T_1 + (length of transition curve).

- Chainage of E_2 = Chainage of E_1 + (length of circular curve).
- Chainage of point of tangency, T_2 = Chainage of E_2 + (length of transition curve).
- Tangent length of combined curve.

$$=(R+S)\tan\frac{\Delta}{2}+\frac{L}{2}\left(1-\frac{S}{5R}\right)$$

Widening of Pavement on Horizontal Curve

In horizontal curves, especially when the radii of the curves are low, it is the practice to widen the pavement width a little extra for safe driving.

1. Reasons for Widening

- Majority of vehicles have the facility to turn only the front wheels and have a rigid wheel base. Jeep like vehicles has a four - wheel drive facility. When vehicles with two - wheel drive facility take a turn to negotiate a horizontal curve, the rear wheels do not follow the same path on that of the front wheels. This is referred to as off - tracking.
- At high speeds when adequate super elevation and lateral friction are not available, there will be some transverse skidding and the rear wheel may take paths on the outside of those traced by the front wheels.
- The drivers may have a tendency, to follow the curved path with larger radius and to have a greater visibility at curve.
- While negotiating an on - coming vehicle or overtaking a horizontal curve there is a psychological tendency for the drivers to adopt a greater clearance between the vehicles.

To take care of these aspects, it has been the practice to widen the pavement on the horizontal curves when the radius of the curve is less than 300 m.

The extra widening needed comprises of two parts, viz:

- Mechanical widening.
- Psychological widening.

2. Mechanical Widening

Mechanical widening is provided to account for off - tracking due to the rigidity of the wheel base. The movement of the vehicle is shown in figure.

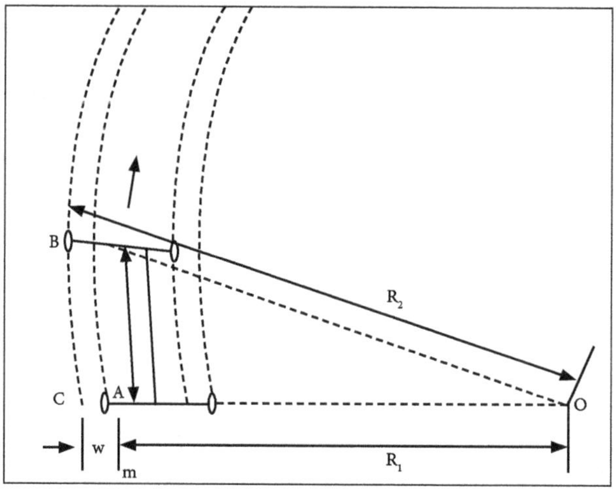

Mechanical Widening on Horizontal curve.

Where,

R_1 - Radius of the path traversed by the outer rear wheel, metres.

R_2 -Radius of the path traversed by the outer front wheel, metres.

W_m - Mechanical widening.

l -Length of the wheel base, metre.

3. Psychological Widening

The psychological widening is mainly provided for greater manoeuvrability of the vehicle. This is particularly important in pavements with more than one lane.

An empirical formula has been recommended by IRC for psychological widening, W_{p8}, as,

$$W_{ps} = \frac{V}{9.5\sqrt{R}}$$

4. Total Widening

Total extra width we provided for a load with n lanes is given as,

$$W_e = \frac{nl^2}{2R} + \frac{V}{9.5\sqrt{R}}$$

Where,

R = Mean radius of the curve,

l = Length of wheel base of longest vehicle (m) (which is taken as 6.0 to 6.1 m for commercial vehicles).

The extra width recommended by the Indian Roads Congress for single and two lane pavements are presented in table.

Table: Extra width of pavement at horizontal curves.

Radius of Curve (m) Extra Width (m)	upto 20	21 to 40	44 to 60	61 to 100	101 to 300	Above 300
Single – Lane	0.9	0.6	0.6	Nil	Nil	Nil
Two - Lane	1.5	1.5	1.2	0.9	0.6	Nil

Note: For multi - lane roads, the pavement widening is calculated by adding half the extra width of two - lane roads to each lane of the multi - lane road.

5. Introduction of Extra Widening

Widening is introduced gradually starting from the tangent point (T.P). If the curve is provided with a transition curve, then it is started from the beginning of the transition curve and provided with uniform rate till the full value of We is reached. At this point the full superelevation is provided. The widening is provided equally on either side of the curve at every cross section up to a value of $W_e/2$.

On horizontal curves without transition curve, two - thirds of the widening is provided at the end of the straight section and one - thirds is provided on the circular curve beyond the tangent point Here widening is provided only on the inside of the curve.

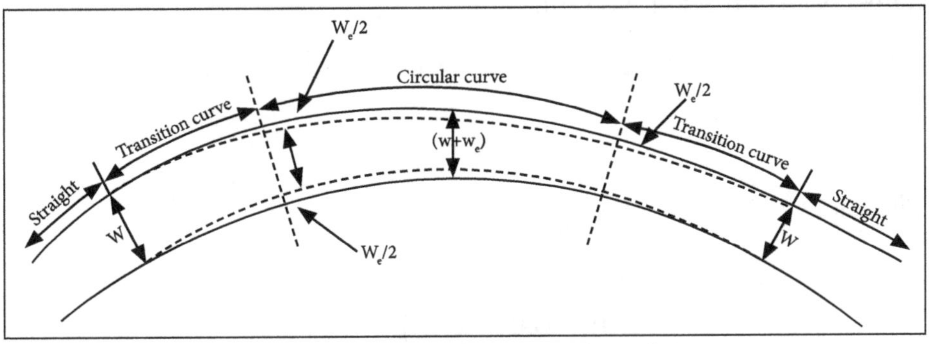

Extra Widening of Pavement on Horizontal Curve.

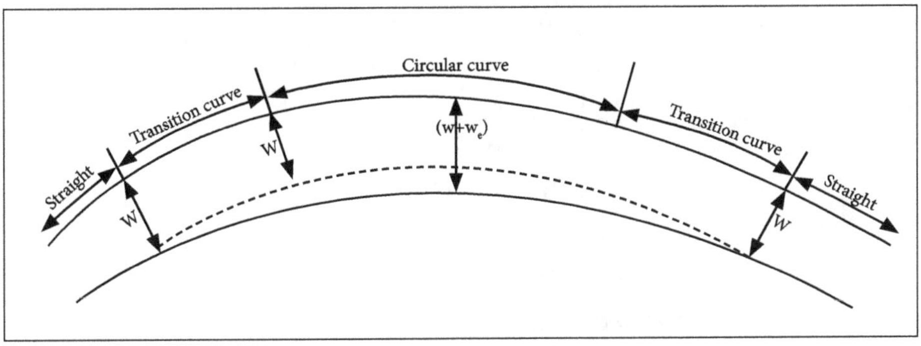

Widening of Pavement on Sharp Curve.

Problem

Design speed for plain terrain, V = 100 kmph.

Number of lanes, n = 2.

Pavement width = 7.0 m.

Wheel base of truck, l = 6.0 m.

Maximum value of superelevation, e = 0.07.

Skid resistance, f = 0.15 with these details. Let us compute the additional width required for a horizontal curve of a state highway in a plain area with a ruling minimum radius.

Solution:

Given:

Design speed for plain terrain, V = 100 kmph.

Number of lanes, n = 2.

Pavement width = 7.0 m.

Wheel base of truck, l = 6.0 m.

Maximum value of superelevation, e = 0.07.

Skid resistance, f = 0.15.

Therefore $R_{ruling} = \frac{V^2}{127(e+f)}$

$$= \frac{100^2}{127(0.07+0.15)}$$

$$= 357.9$$

$$= 360 \text{ m}$$

Additional width, $W_e = \frac{nl^2}{2R} + \frac{V}{9.5\sqrt{R}}$

$$= \frac{2\times6^2}{2\times360} + \frac{100}{9.5\sqrt{360}}$$

$$= 0.10 + 0.56$$

$$= 0.66 \text{ m}$$

$$\text{Total pavement width} = 7.0 + 0.66$$
$$= 7.66\text{m}$$

2.5 Vertical Curves

In a highway a change in the rate of grade or direction may cause the vehicle a sudden impact. For a smooth and safe running of vehicles, the change in gradient or direction is smoothened by a curve called a vertical curve. Vertical curve is used to smooth out intersecting grade or change in direction of roads.

A parabolic curve provides a smooth and most comfortable ride as the rate of change of grade in a parabola with a vertical axis is uniform throughout. The vertical curves adopted in highway may be classified as:

- Summit curve or Crest curve with convexity.
- Valley or Sag curve with concavity upwards.

Summit Curves

A curve with convexity upwards is called a summit curve. This occurs when,

- An ascending gradient meets another ascending gradient.
- An ascending gradient intersects a descending gradient.
- A descending gradient meets another descending gradient.
- An ascending gradient meets a horizontal.

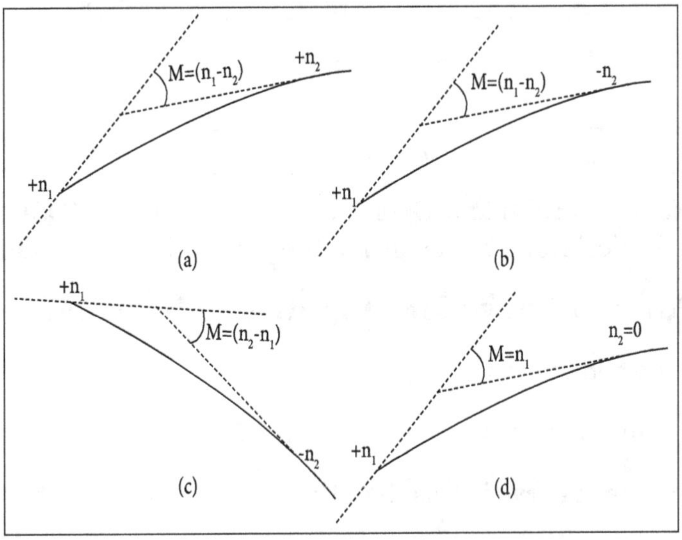

Types of Summit Curves.

1. Suitability of Curve for Summit Curve

Circular curve is ideal for a summit curve as the sight distance available throughout the curve is constant. A transition curve is undesirable as the radius of curvature changes and hence the sight distance will vary from point to point. The deviation angles in vertical curves are very small.

As a result, tangent points of a circle and a parabola are congruent. A parabola is very easy for arithmetical manipulation for computing ordinates. Use of simple parabola as summit curve is found to give good riding comfort.

2. Length of Summit Curve

The equation for parabolic summit curve is given by,

$$y = ax^2$$

Where, $a = \frac{N}{2L}$, N = Deviation angle and L = Length of the curve.

As the summit curves are long and flat, the length of the curve is taken as its horizontal projection, e.g., L = A Jas shown in figure:

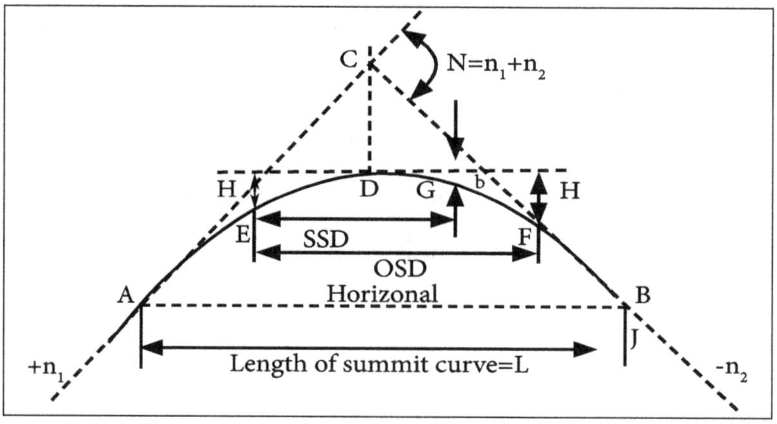

Details of a Summit Curve.

Stopping Sight Distance (SSD) and Overtaking Sight Distance (OSD) should be separately considered for designing the length of the parabolic summit curve.

3. Length of Summit Curve for Stopping Sight Distance

Two cases have to be considered:

- Length of summit curve is greater than the Stopping Sight Distance (L > SSD).
- Length of summit curve is less than the Stopping Sight Distance (L < SSD).
- L> SSD.

The general equation for length L of the parabolic curve is given as,

$$L = \frac{NS^2}{\left(\sqrt{2H} + \sqrt{2h}\right)^2}$$

Where,

L=Length of summit curve, metres.

S=Stopping sight distance.

N=Deviation angle (equal to algebraic difference in grades).

H=Height of eye level of driver above the road surface, metres.

h=Height of subject above the road surface, metres.

The value of H, is 1.2 m in India and h is 0.15 m as per IRC. Substituting these values in the above Equation, we get,

$$L = \frac{NS^2}{4.4}$$

- L<SSD

For the condition that L is less than SSD is given as,

$$L = 2S - \frac{\left(\sqrt{2H} + \sqrt{2h}\right)^2}{N}$$

In order to find the length of summit curve, the SSD is calculated or taken from table.

Length of summit curve from above two equations by trial is used to get the required length.

4. Length of Summit Curve for Safe Overtaking Sight Distance or Intermediate Sight Distance

Here again two cases are to be considered:

- (L>S) and
- (L<S).

Where,

L - Length of the summit curve.

S - Overtaking sight distance or intermediate sight distance.

- L> S

The general equation is applicable in this case also.

In this case h = H = 1.2 m. Substituting h = H, we get,

$$L = \frac{NS^2}{8H}$$

$$\text{i.e., } L = \frac{NS^2}{8 \times 1.2} = \frac{NS^2}{9.6}$$

- L < S

Here again substituting h = H = 1.2 m in the general equation.

$$L = 2S - \frac{9.6}{N}$$

In both the cases the minimum radius of the curve R is calculated from the relation R = L/N.

When the deviation angle is small, the length of summit curve works out less than the sight distance.

When the deviation angle is very small, a negative value for length of summit curve may be obtained which indicates that there is no sight restriction at the summit curve.

Minimum length of vertical curve for different speeds and for the maximum grade change which do not need vertical curve is shown in table.

Table: Requirement of grade change.

Design Speed, (kmph)	Maximum Grade Change (Percentage) not Requiring A Vertical Curve	Maximum Length of Vertical Curve (Metres)
35	1.5	15
40	1.2	20
50	1.0	30
65	0.8	40
80	0.6	50
100	0.5	60

The highest point on the summit curve is a distance Ln^N from the tangent point on the first grade n_1.

Problem

1. Let us consider a road with an ascending gradient of 1 in 120 is followed by a descending gradient of 1 in 100 and the design speed on the road is 100 kmph and the overtaking sight distance is 510 m. calculate the length of the summit curve.

Solution:

Given:

Ascending gradient of 1 in 120,

Descending gradient of 1 in 100,

Design speed on the road is 100 kmph,

Overtaking sight distance is 510 m.

As this value is less than OSD of 510 m, assume L is less than OSD Then for L < OSD,

$$L = 2S - \frac{9.6}{N}$$

$$= 2 \times 510 - \frac{9.6 \times 600}{11}$$

$$= 1020 - 523.6$$

$$= 496.4 \text{ m}$$

This value is less than the OSD of 510 m.

Length of summit curve = 496.4 m.

Valley Curves

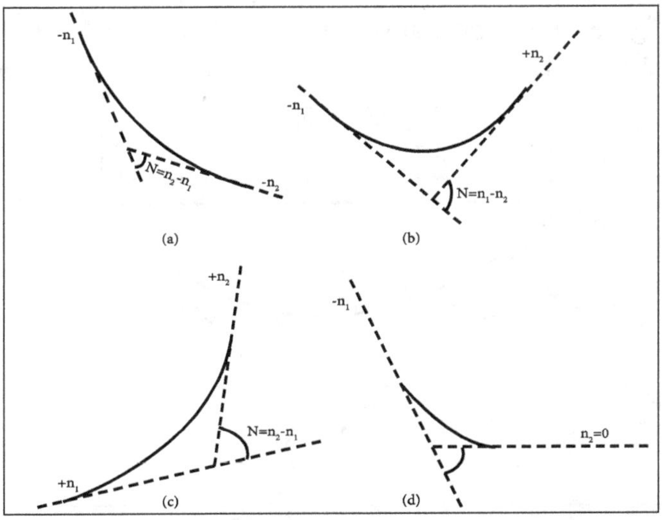

Types of Valley Curves.

A vertical curve, concave upwards, is called a valley or sag curve. This occurs when,

- A descending gradient meets another descending gradient.
- A descending gradient intersects an ascending gradient.
- An ascending gradient meets another ascending gradient.
- A descending gradient meets a horizontal.

1. Suitability of Curve for Valley Curve

In a valley curve, a large component of the centrifugal force will act vertically downwards. It exerts extra pressure on the tyres and springs of the vehicle. Sudden entry of the vehicle from a straight path (R is infinity) to a curved path of fixed curvature causes a shock and discomfort. Centrifugal force varies with the square of the speed and therefore may cause trouble if the roads are not planned.

It is necessary that the centrifugal force should be gradually applied. To satisfy this, the property of the curve should be such that its radius of curvature decreases gradually as its length increases. Hence transition curve is a most suitable of the three transition curves, viz., spiral, cubic parabola and the lamniscate, Cubic parabola satisfies the requirements for the valley curve.

2. Length of Valley Curve

The length of a valley curve is designed on the basis of two criteria:

- The allowable rate of change of centrifugal acceleration is 0.6 m/sec^2.
- The head light sight distance.

High value of length based on the above two criteria is adopted. Second criterion also gives the higher value. Figure shows the transition curve, ABC with AB and BC are two equal transition curves with, $L_s = L/2$, having the minimum radius R at the common point B.

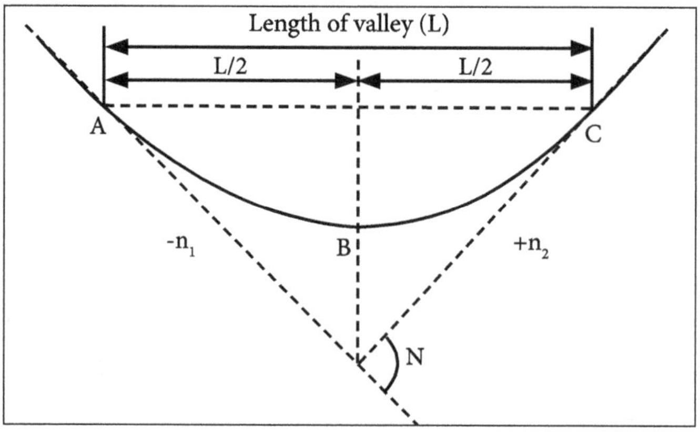

Details of Valley or Sag Curve.

For Comfort Condition

The length of transition curve L for comfort condition is given as,

$$L_s = \frac{v^3}{CR}$$

Value of R at length $L_s = L_s/N = \frac{\quad}{2N}$.

Substituting in the above equation,

$$L_s = \frac{v^3}{CL_s} N \text{ Or, } L_s^2 = \frac{Nv^3}{C}$$

$$\text{i.e., } L_s = \sqrt{\left[\frac{Nv^3}{C}\right]}$$

$$\text{i.e., } L = 2L_s = 2\sqrt{\left[\frac{Nv^3}{C}\right]}$$

Where,

v = Speed in m/sec.

C = Allowable rate of change of centrifugal acceleration.

Taking C = 0.6 m/sec² and taking speed in kmph.

Then,

$$L_s^2 = \frac{Nv^3}{0.6 \times 3.6^3}$$

$$\text{i.e., } L_s = 0.19\sqrt{\left(Nv^3\right)}$$

$$\text{i.e., } L = 2L_s 0.38\sqrt{NV^3}$$

The minimum radius (R in metres) of the valley curve for cubic parabola is given as,

$$R = \frac{L_s}{N} = \frac{L}{2N}$$

For Head-Light Sight Distance Condition

The length of valley curve has to be determined for the two conditions:

- When the total length of valley curve L is greater than the stopping sight distance.

- When L is less than SSD, as given below.

1. L > SSD

The length of valley curve L is assumed to be greater than the head-light sight distance which should be equal to SSD.

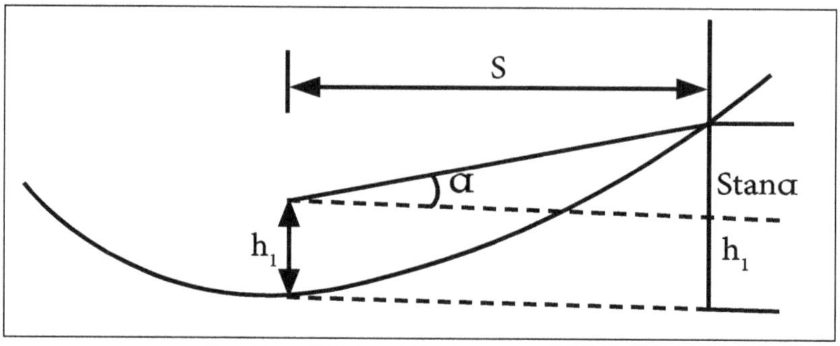

Details of head-light sight distance when L > S.

The sight distance will be minimum when the vehicle is at the lowest point on the sag curve. Assuming the valley curve as a parabola, then y =ax² and

$$h_1 + S\tan\alpha = \alpha S^2 = \frac{NS^2}{2L} \text{ (Since, a = N/2L)}$$

$$\text{Then, } L = \frac{NS^2}{2h_1 + 2S\tan\alpha}$$

Where,

h_1 = Height of the head light,

α = Beam of head-light inclination,

N = Deviation angle $(n_1 + n_2)$ with slopes $-n_1$ and $+n_2$,

Substituting for $a = 1°$, $h_1 = 0.75$ m,

$$L = \frac{NS^2}{(1.5 + 0.035S)}$$

2. L< SSD

Let the vehicle be at the start of the valley curve or at the tangent point, TP, for minimum sight distance. Then,

Substituting for $a = 1°$, $h_1 = 0.75$ m in the Equation for,

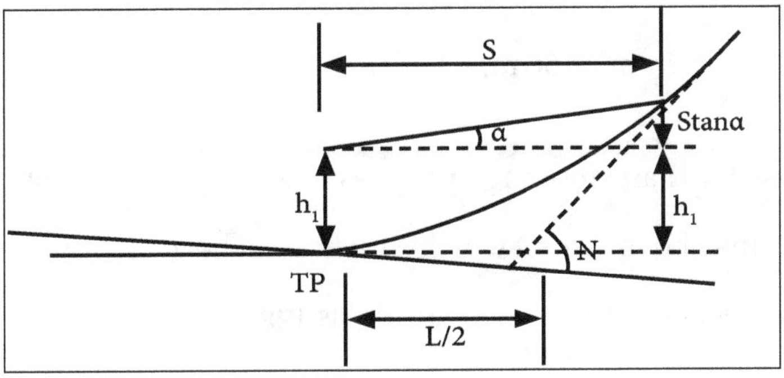

Head lights right distance when L<S.

To provide the cross drainage facility, the lowest point on the valley curve should be located. If the gradients are equal, then the lowest point will lie on the bisector of the angle between the curves. On unequal grades the lowest point lies on the side of the flatter gradient.

Problem

1. Let us consider a descending gradient of 1 in 25 meets an ascending gradient of 1 in 45 on a district highway in plain terrain. Compute the length of the sag curve which should provide safe driving at night.

Solution:

Given:

Descending gradient of 1 in 25 meets an ascending gradient of 1 in 45.

$$N = \frac{1}{25} - \left(-\frac{1}{45}\right) = 0.04 + 0.022$$
$$= 0.062$$

For a District Highway in plain terrain, the stopping distance = 120 m.

(i) Length of sag curve $= 0.38\left(NV^3\right)^{1/2}$

$$= 0.38\left(0.062 \times 80^3\right)^{1/2}$$

$$= 67.70 \text{ m.}$$

(ii) Assuming L > S, then

$$L = \frac{NS^2}{1.5 + 0.035S} = \frac{0.062 \times 120^2}{1.5 + 0.035 \times 120}$$

$$= \frac{892.8}{5.7} = 156.63\,\text{m} = 160\,\text{m}.$$

As this value is greater than stopping distance of 120 m this assumption is correct.

Hence L = 160 m, may be provided which satisfies the head-light requirements also.

Hence the length of valley curve may be taken as 132 m.

2.6 Gradients and Special Consideration for Hill Roads

Gradient

It is the rate of rise or fall of a road level along its length. It is expressed as a percentage rise or fall or a rate of rise or fall with respect to horizontal distance.

A gradient of 1 in 20 or 5% represents that there is an ascending or descending of road profile by one meter for every twenty metres. As the gradient on the road is not very steep the length along the road is taken as a horizontal length.

Grades are expressed as the horizontal distance along the length of the road. Longitudinal grades are almost permanent features of the highway. This is the case in urban and built up areas where no changes can be incorporated easily.

The design of longitudinal grades depends on following factors:

- Anticipated traffic.
- Site conditions:
  - Drainage.
    - Appearance.
    - Approach to adjacent property.
    - Safety aspects.
- Railway - intersections.
- Bridges and culverts.
- Design speed.
- Sighting line and sight distance.

1. Types of Gradient

Gradients are classified as:

- Ruling gradient,
- Limiting gradient,
- Exceptional gradient,
- Minimum gradient.

Ruling Gradient

It is the desirable upper limit of gradient adopted in the normal course of design. Adoption of ruling gradient should balance the cutting and filling of earth work which will give an economical design. Different factors which are to be considered in the choice of ruling gradient are:

- Type of terrain,
- The length of the grade,
- The speed,
- Pulling capacity of vehicles,
- The presence of horizontal stretches of road.

Indian Roads Congress has recommended gradients for roads in different terrains which are specified in the table below.

Table: Gradients for roads in different terrains (CRC).

Type of Terrain	Rolling Gradient	Limiting Gradient	Exceptional Gradient
Plain or Rolling	3.3% (1 in 30)	5.0 % (1 in 20)	6.7 % (1 in 15)
Mountanious terrain and steep terrain having elevation more than 300 m above mean sea level	5.0 % (1 in 20)	6.0 % (1 in 16.7)	7.0 % (1 in 14.3)
Steep terrain up to 3000m height above mean sea level	6.0% (1 in 16.7)	7.0 % (1 in 14.3)	8.0 %(1 in 12.5)

Limiting Gradient

It is provided when topography is not permitting to provide the ruling gradient, due to economy. Length of continuous grade of stretch of road steeper than ruling value should be limited. It happens in rolling and hilly terrains to go in for limiting gradient but it has to be separated by straight stretches or road with easier grade frequently.

Exceptional Gradient

In some ground conditions it will be inevitable to provide a gradient steeper than limiting gradients such gradients are referred to as exceptional gradient. Such situations may arise in approaches to causeways, near hair - pin bends, etc. In a hilly terrain short stretches of grades, steeper than the maximum gradient may have to be resorted to reach an obligatory point on a fixed route.

The exceptional gradients have several adverse effects, viz., more fuel consumption, more friction losses, reduces engine efficiency, increased wear of road surface, etc. Hence it should be adopted only in very essential field conditions.

Minimum Gradient

A road with less gradient or level may not be in a position to drain easily. The surface water may be drained to the side drain due to the camber. A longitudinal slope is needed to drain the water on the surface and from the side drains.

It is essential to have a certain minimum gradient on roads from drainage point of view provided topography permits. The minimum gradient depends on the factors like rainfall, runoff, type of soil, topography and other site conditions.

2. Grade Compensation

In addition to gradient if there is horizontal curve then there will be added resistance to traction.

In such cases, the total resistance due to grade and curve should not exceed the resistance due to the maximum value of the gradient specified.

For design purposes, the maximum gradient may be considered as that of ruling gradient. In some special cases the limited gradient may be used.

The gradient should be reduced in situations where sharp horizontal curves are present so as to compensate for the loss of tractive effort due to the curve. This reduction in gradient at the horizontal curve is called the grade compensation.

It is provided to offset the extra tractive effort involved at the curve.

Grade compensation (%) = $\frac{30+R}{R}$ subject to maximum value of 7.5 / R,

Where,

R is the radius of the circular curve in metres.

As per IRC, there is no need for grade compensation for gradients flatter than 4.0 %.

Special Consideration for Hill Roads

Geometric designs of hill roads are different from those in plains. Main reasons for the difference are the topography and other problems. The design speed should be different. Recommendations of Indian Roads Congress (IRC) for hill roads are discussed below.

Design Speed

For hilly regions design speeds as per IRC recommendations are as follows.

Table: Design speed in Hilly Region.

Design Speed km/hr	National Highway and State Highway	Major District Roads	OtherRoads	Village Roads
Minimum	50	40	30	25
Maximum	60	50	40	35

Cross Section Elements of Hill Roads

As a general rule, geometric features of a highway except cross - sectional elements do not lead to stage construction. In the case of hill roads, features like grade and curvature cannot be changed at a later date easily. It may be very expensive and may not be possible at all.

A careful design should be made regarding alignment and to some extent cross - section elements. Development to cross section should be decided very carefully as it involves various additional construction like retaining walls, breast walls, catch water drains etc.

1. Formation Width

Formation widths and pavement widths as per IRC are given in table.

Table: Formation Widths.

Type of Highway	Land Width		Formation Width in Metres (Excluding Side Drain and Parapets)	Carriage way Metres
	Normal	Exceptional		
National of Highway	24	18	8.80 to 6.25	7.0
State Highway	24	18	8.80 to 6.25	3.72
Major District Road	18	15	4.75	3.75
Other District Road	15	12	4.75	3.75
Village Road	9	9	4.00	3.00

2. Sight Distance

The Stopping Sight Distance is calculated from the relation,

$$SSD = 0.278\,Vt + \frac{V^2}{254f}$$

Where,

V = Design speed, kmph.

t = Reaction time, taken as 3 seconds.

f = Coefficient of friction, assumed as 0.4.

Safe Stopping Sight Distances for various speeds given by IRC are given in Table.

Table: Safe Stopping Sight Distance.

Speed, kmph	20	25	30	40	50
SSD, m	20	30	35	50	70

Overtaking Sight Distance is calculated from the relation:

$$OSD = 5.556\,V_b + 2\,S + 0.278\,T\,V_b + 0.278\,V$$

Where,

V = Speed of overtaking vehicle, kmph.

V_b = Speed of overtaken vehicle kmph.

= (V - 16) km/ph

S = Spacing of moving vehicles

= $(0.2\,V_b + 6)$ m.

T = Overtaking time, sec.

$= \sqrt{\frac{14.4}{A}}$ secs.

A = Acceleration in kmph/sec.

Minimum Overtaking Sight Distance for various speeds as recommended by IRC is given in table.

Table: Minimum OSD.

Speed, kmph	30	40	50
OSD, m	90	145	210

3. Radius of Horizontal Curve

The minimum radius of horizontal curves in hill roads is calculated from the relation.

$$R_{min} = \frac{0.008V^2}{e+f}$$

Where,

e = Super elevation.

f = Lateral friction factor = 0.15.

Minimum radius recommended for various class of hill roads as per IRC are given in table.

Table: Minimum radius of horizontal curves.

Type of Road	Minimum Radius in Metre			
	Mountainous		Steep Hilly	
	No snow	Snow bound	No snow	Snow bound
National Highway and State Highways	50	60	30	33
Major District Roads	30	33	14	15
Other District Roads	20	23	14	15
Village Roads	14	15	14	15

Reverse curves are designed for a minimum radius of 30 m for the compound curves and a straight distance of 9 m between their transition ends.

4. Super Elevation

Super elevation to be provided in horizontal curves of hill roads may be computed from the relation given below for a given minimum radius of horizontal curve.

$$e = \frac{V^2}{225R}$$

IRC has recommended a maximum value of super-elevation as 1 in 15 and 1 in 10 for hill roads of snow bound.

5. Widening of Curve

Extra width of carriageway, W_e at horizontal curve is given by the relation as,

$$W_e = \frac{18n}{R} + \frac{0.1V}{\sqrt{R}}$$

Where,

'n' is the number of lanes.

Extra widening as per IRC recommendation is given in table.

Table: Extra widening requirement.

Radius of Curve, Metres	14 to 20	20 to 30	30 to 60	60 to 150	Above 150
Single lane: Extra Widening Metres	1.5	1.20	0.90	0.6	Nil
Radius of Curve, Metres	30 to 40	40 to 60	60 to 100	100 to 150	Above 150
Two Lane: Extra Widening, Metres	1.5	1.20	0.90	0.60	Nil

Extra formation width should also be provided at horizontal curves.

Minimum of 0.5 m shoulder width should be provided on either side of the carriage way.

2.7 Hairpin Bends

In hill roads it is often needed to provide hair-pin bends depending on the topography. The geometric standards adopted for the conventional roads cannot be adopted for hairpin bends. Design of such hair-pin bends needs special consideration.

The Indian Standard Practice (1944) for the design of hair-pin bends are given below:

- Minimum design speed 20 km/hr.
- Minimum width at apex:
    - National and state Highways.
        - Double lane 11.5 m.
        - Single lane 9.0 m.
    - Major District Roads and Other District Roads 7.5 m.
    - Village Roads 6.5 m.
- Minimum radius of inner curve 14.0 m.
- Minimum length of transition 15.0 m.
- Gradient.

    ◦ Maximum 1 in 40.
    ◦ Minimum 1 in 200.

- Superelevation 1 in 10.
- Distance between two consecutive hair-pin bends should be a minimum of 60 m.
- Approach gradient should not be steeper than 5%.
- Straight length (60 m), if possible, should be introduced between two successive hair-pin bends.

A typical hair-pin bend is shown in figure:

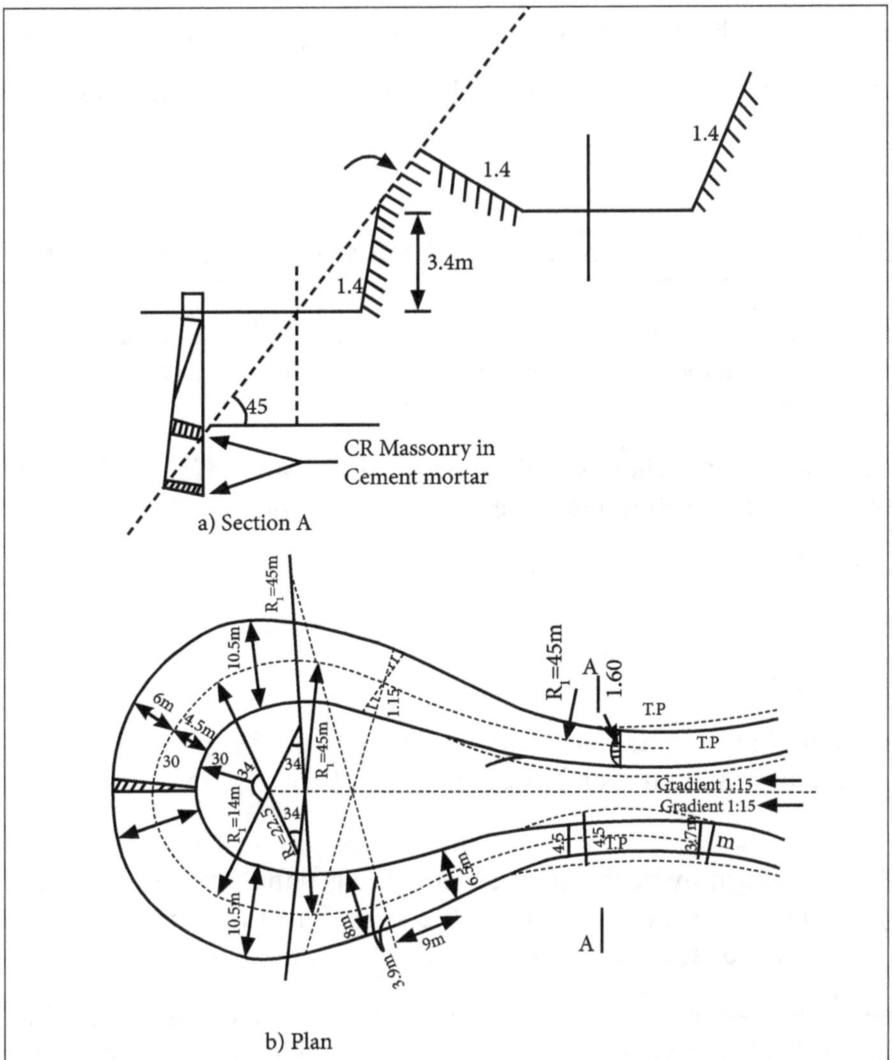

Typical Hair-Pin Bend.

The minimum norms which may be adopted in the design of hair-pin bends are as follows:

- Minimum radius = 14 m.
- Design speed at the curve = 20 km/hr.
- Maximum super elevation = 1 in 10 and for light traffic 1 in 15.
- Width of carriage way at the apex of the curve should be 11.5 m for two-lane and 9.0 m for single lane National Highway and State Highway and for District Roads 7.5 m.
- Maximum and minimum gradients are 1 in 40 and 1 in 200 respectively at curves.
- Approach gradient should not be steeper than 5 % for 40 metre (60 m).
- Straight length (60 m) roads should be introduced between two successive hair-pin bends.

2.8 Lateral and Vertical Clearance at Underpasses

Highways may have to cross a roadway or railway. At this junction grade separation structures are constructed. Structures may be T-beam Bridge, arch bridge, prestressed concrete bridge, etc.

For conventional vehicles, there should be vertical clearance of at least 4.3 m. For double-decker vehicles the vertical clearance should be 5.2.m.

Grade separated intersections are classified as:

- Over-pass,
- Under-pass.

When the major highway is taken above by raising its profile above the general ground level by embarkment and an over-bridge across another highway or railway, it is called as over-pass.

If the highway is taken by depressing it below the ground level to cross another road or railway by means of an under bridge, it is known as under-pass. The details of an underpass for cross roads in an expressway are shown in figure.

The choice of an over-pass or an under-pass depends on various conditions such as topography, vertical alignment, drainage, economy, aesthetic features and preferential aspects for one of the highways.

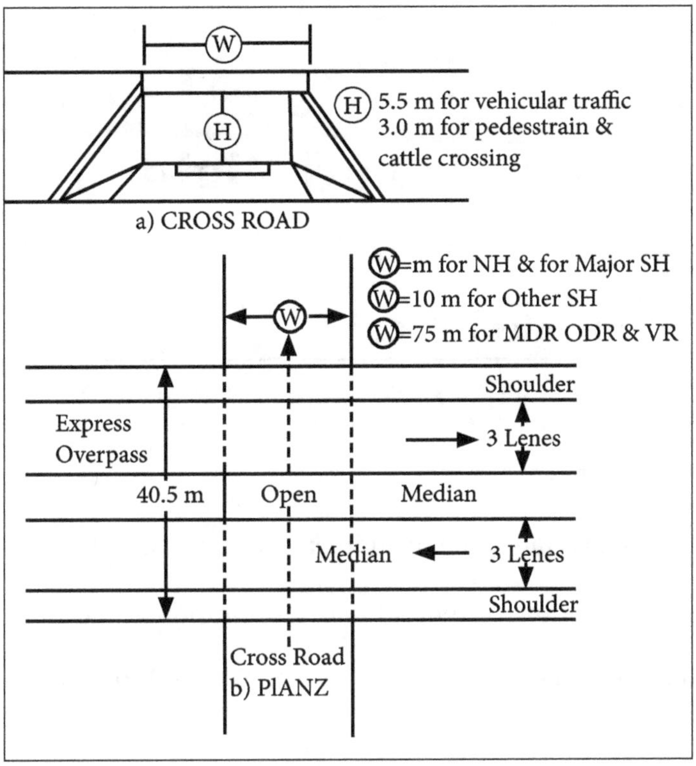

Underpass for cross-roads in an Express way.

Merits and Demerits of Under-pass

1. Merits

- Adequate warning is provided to the traffic in advance due to the presence of an under-pass which can be seen from distance.
- Useful for the turning traffic.

2. Demerits

- Possibility of drainage problem at the under-pass where the ground water level rises high.
- During water-logging problem in monsoons, continuous pumping of water may be needed.
- There is some restriction of the vertical sight distance even at the valley curve near the under-pass.
- Unless the clearance is sufficiently large, there is a feeling of restriction to the traffic.
- No possibility of stage construction for the bridge structure at the under-pass.

3

Design of Flexible and Rigid Pavements

3.1 Design Principles

Design Factors

Pavement design comprises of two parts, viz:

- Mix design of materials to be used in each component layer.
- Thickness design of the pavement and the component layer.

The mix design is made depending on the type of material to be used for the load and other conditions.

Design Wheel Load

Thickness design of pavement primarily depends upon the design wheel load.

The design wheel load depends on the following:

- Maximum wheel load.
- Contact pressure.
- Equivalent single wheel load.

Maximum Wheel Load

In order to know the way by which load is distributed, the wheel load configuration should be known. Figure shows the typical wheel load configuration of a tractor - trailer unit of a heavy duty vehicle. Before deciding the wheel load configuration to be selected, the legal axis load specified in the area should be taken into account. Indian Roads Congress has specified the maximum legal load as 8170 kg and maximum single wheel load as 4085 kg.

The configuration or shape of wheel load gives an idea how the load of a vehicle acts on the surface of the pavement. The vertical stress σ_z at depth z, can be computed using Boussinesq's equation under a uniformly distributed circular load as,

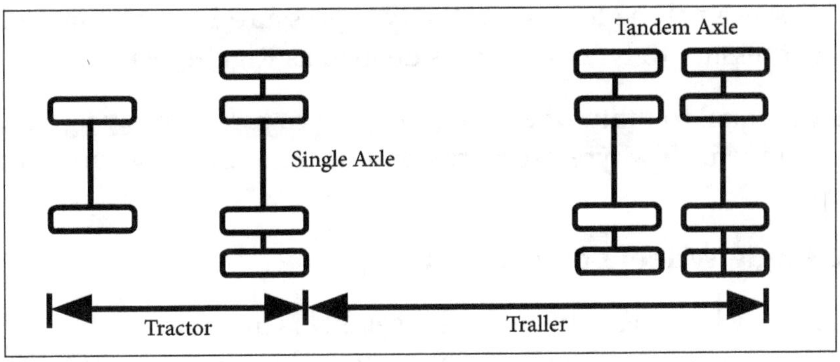

Wheel configuration of tractor - trailer unit.

$$\sigma_z = p\left[1 - \frac{z^3}{\left(a^2 + z^2\right)^{3/2}}\right]$$

Where,

σ_z = Vertical stress at depth z,

p = Surface pressure,

Z = Depth at which σ_z is computed,

a = Radius of loaded area.

Using the above equation the variation of vertical stress with respect to depth can be computed. The distribution will have the highest value near the surface and decreases with increase in depth.

Contact Pressure

The wheel load is assumed to be distributed over a circular area. Distributed load is the tyre pressure which is differently referred to as inflation pressure or contact pressure. Tyre pressure and inflation pressure mean exactly the same. Contact pressure is given by the relationship,

Contact pressure, p = L/ A

Where,

p = Contact pressure,

L = Load on wheel,

A = Contact area or area of imprint.

The contact pressure is found to be more than tyre pressure when the tyre pressure is less than 7 kg/cm^2.

The contact pressure will be less when the tyre pressure is more than 7 kg/cm². The ratio of contact pressure to tyre pressure is defined as Rigidity Factor.

Rigidity factor is equal to unity when the average tyre pressure is 7 kg/cm². This value is higher than unity for low tyre pressure and less than unity for tyre pressure higher than 7 kg/cm².

Equivalent Single Wheel Load (ESWL)

To have maximum wheel load, dual wheel assembly is provided to the rear axles of the load vehicles. Because of this, the load due to both the wheels is not to be transferred to the pavement. But there will be overlap pressure after a certain depth. Actual effect is in between a single wheel load and double the load carried by any one wheel. Stress overlap is presented in figure:

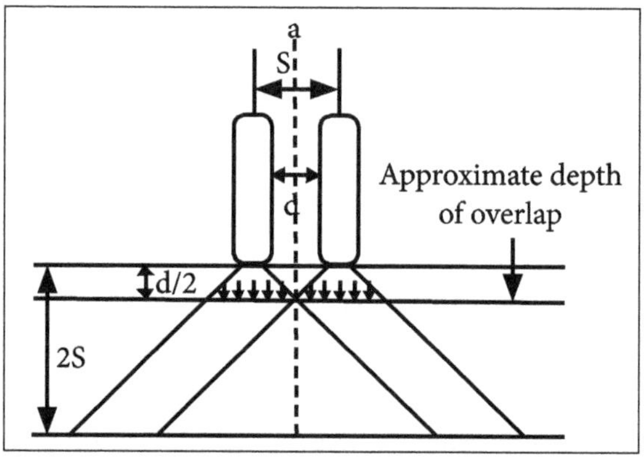

Stress overlap due to dual wheels.

Assume that up to a depth of d/2, the loads act independently beyond which the stresses overlap. The area of overlap becomes more beyond a depth of 2 S. Hence consider that the total stress due to the dual wheels at any depth greater than 2 S, is equivalent to a single wheel load of 2 P magnitude. This stress due to 2 P is to be slightly greater than the dual wheel assembly which is on the safe side. This Equivalent Single Wheel Load (ESWL) can be determined by equivalent deflection or equivalent stress criterion.

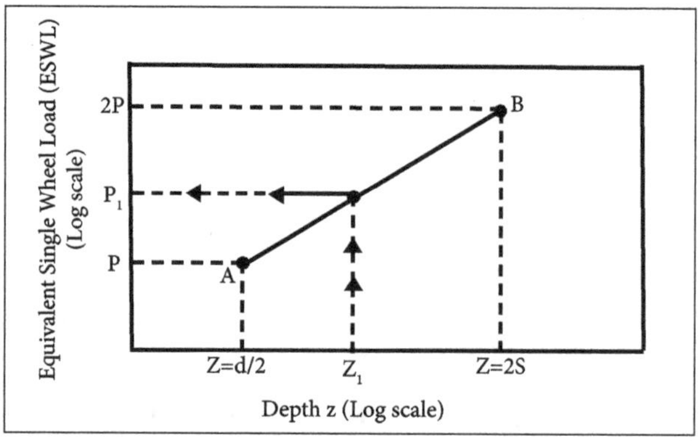

Graphical representation of ESWL.

Line AB represents the locus of points where any single wheel load is equivalent to a certain set of dual wheels. In order to use the graph, for an assumed thickness of pavement and ESWL is got from the graph. This ESWL is used in the design calculations and the thickness of pavement is obtained. If the thickness and assumed thickness are same, then the ESWL assumed is correct. If the design is not repeated and by trial and error, the correct thickness of pavement is obtained. This is valid for the given wheel load configuration. For different wheel load assembly different ESWL plots may be made.

3.2 Pavement Components and their Role

Pavements components are as follows:

- Soil,
- Sub grade,
- Sub base,
- Base course,
- Surface course.

Requirements of Pavement Components

Selection of pavement components for the selected pavement such as type of pavement, base course, sub-base and surface course is governed by the following factors:

- Type of subgrade soil.
- Drainage conditions.
- Type of traffic and volume of traffic.
- Availability of construction material along the alignment or within a reasonable distance from the site.
- Availability of plant and equipment.
- Climatic conditions.
- Altitude at which the road is proposed to be constructed.
- Time available to complete the project.
- Availability of funds for construction.
- Fund availability for maintenance.

The selection of pavement type depends on the factors mentioned above. Out of them, the important factor is the traffic volume or intensity of traffic.

The type of road which could be considered based on traffic is listed below:

Traffic Intensity tonnes Per Day	Type of Road
30 to 200	Low cost
up to 500	Water Bound Macadam
up to 1500	Bituminous
up to 2000	Cement concrete

These roads depend on the thickness of pavement section, type of soil subgrade and the desired service life of the road. Each pavement has its own life span. There should be provision to improve the pavement structure in stages to match with increase in road traffic. In order to accommodate this aspect, stage construction technique should be adopted.

3.3 Design Practice for Flexible and Rigid Pavements (IRC Methods Only)

Flexible pavements consist of different thickness of layers of different materials. Due to the wheel loads, the maximum intensity of stress occurs in the top layer of the pavement. Magnitude of stress decreases with depth. Superior quality of materials is used in the top layers followed by less quality materials.

In the design process, it is the basic requirement that under the application of the load none of the layer is over stressed. No section of the pavement undergoes excessive deformation which may result in localized depression or settlement. There is no rational design method for flexible pavement. The flexible pavement design methods may be classified for distinct groups:

1. Empirical Methods

These are based on soil classification, physical or strength parameters of soils and other factors such as climate and moisture.

2. Semi-empirical Methods

These are methods based on stress strain functions, performance and serviceability concepts.

3. Theoretical Methods

These are based on theoretical analysis and mathematical computations considering stress-strain behaviour of soils.

Various methods which are in use are discussed below.

Group Index Method

This method of flexible pavement design is based on the soil properties like,

- Particle size distribution.
- Liquid limit.
- Plasticity index.

Group index is claimed to be an inverse measure of the thickness of pavement. The higher is the group index, the lower is the strength and greater is the thickness requirement for a given traffic volume.

According to this method, subgrades have been divided into five based on group index:

Subgrade Quality	Value of Group Index
Excellent	0
Good	0 to 1
Fair	2 to 4
Poor	5 to 9
Very poor	10 to 20

Traffic is divided into three groups:

Total Volume (Commercial Vehicles)	Number of Vehicles Per Day
Light	Less than 50
Medium	50 to 300
Heavy	Over 300

In order to design the pavement thickness by this method, the following steps are followed:

Step 1 - The value of Group Index is found using the details of the soil.

Step 2 - The anticipated traffic is estimated and classified as light, medium or heavy as indicated in figure.

Step 3 - The appropriate design curve is chosen from figure and the total thickness of pavement is found for the corresponding group index.

Step 4 - The thickness of sub-base or base is found from appropriate curve, to find the thickness.

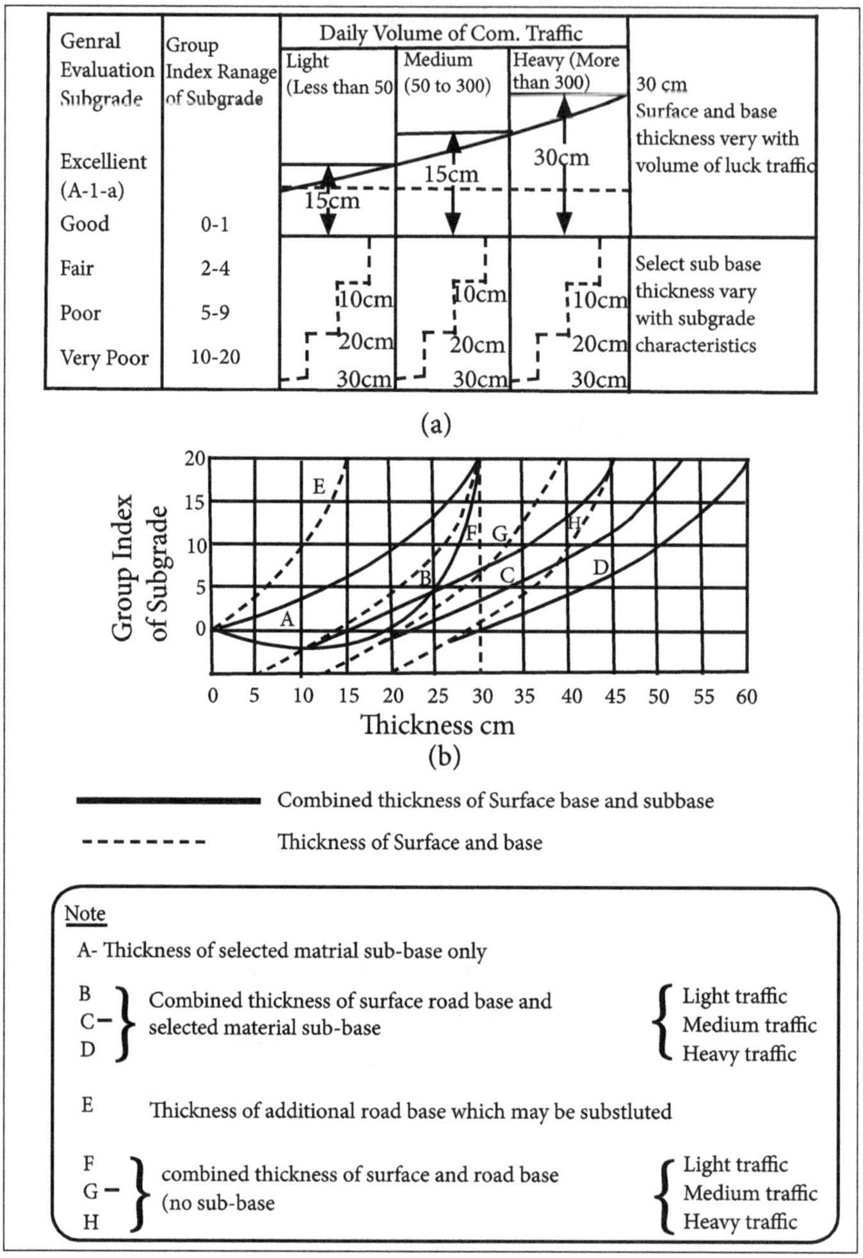

Design chart by Group Index value.

The method is only empirical and considers only the soil physical properties and not the strength and compressibility characteristics of the soil.

Recommendations of Indian Roads Congress

Indian Roads Congress (IRC: 37-1970) has recommended some important aspects to be considered while using the design chart. Following are the recommendations:

- The specimen to be tested for CBR should be re-moulded specimen prepared preferably by static compaction wherever possible or by dynamic compaction. The standard test procedure should be strictly followed.

- In-situ test specimens are not recommended.
- For new roads the subgrade soil specimen should be compacted to proctor density at OMC. If the compaction equipment is not available in the field, the specimen may be compacted to the expected field density.
- For the existing roads the specimen should be compacted to field density of subgrade soil at water content equal to OMC or field moisture content.
- For all new constructions the specimen should be soaked for four days prior to testing, this condition is not mandatory for arid climatic regions or regions with annual rainfall less than 50 cm or the water table is very deep or when thickness of impermeable bituminous surfacing is provided.
- At least three specimens should be tested with identical specimens. If the variation is maximum beyond the norms, then average of six specimens CBR values should be taken.

The specified limits of maximum variation in CBR values are,

- 3 % for CBR value up to 10 %.
- 5 % for CBR value 10 to 30 %.
- 10 % for CBR value 30 to 60 %.
- The top 50 cm of subgrade should be compacted with density equal to 95 to 100 % of proctor density.
- Keeping in view the existing traffic and the anticipated growth in traffic should be calculated for at least 10 years of life period.

The following formula may be used:

$$A = P[1+r]^{n+10}$$

Where,

A = Number of heavy vehicles per day for design (laden weight >3 tonnes).

P = Number of heavy vehicles per day at last count.

r = Annual rate of increase of heavy vehicles.

n = Number of years between the last count and the year of completion of construction.

The value of P has to be found for seven-day heavy vehicles obtained from 24-hours count.

If a reasonable value of r is not available a value of 7.5 % may be assumed for rural roads:

- The traffic obtained from the above equation has to be used in choosing the appropriate design curve (A to G).
- The design thickness corresponding to a single axle load up to 8200 kg and tandem axle loads up to 14500kg is adopted.
- Substandard sub-bases with substantial proportion of aggregates of size above 20 mm should not be used in design.
- Thin layers of wearing course such as surface dressing or open graded premixed carpet up to 2.5 cm thickness should not be counted towards total thickness as these materials do not contribute to the structural capacity of the pavement.

3.4 Embankments

It is constructed when the grade line of a highway is required to be raised above the existing ground level. Necessity for raising of ground level for construction of highway may arise due to the following reasons:

- In order to make the subgrade level above the high ground water level.
- In order to prevent the capillary water to enter the subgrade by capillary action or surface water through permeation.
- In order to satisfy the vertical alignment of the designed highway.

Design and Construction of Embankment

Design and construction of embankment comprises of:

- Deciding the height of embankment.
- Finalizing side slopes and their short and long term stability.
- Embankment materials.
- Compaction requirement and compacting with suitable equipment.
- Check for settlement and overall stability.

Figure shows the cross section of a typical embankment:

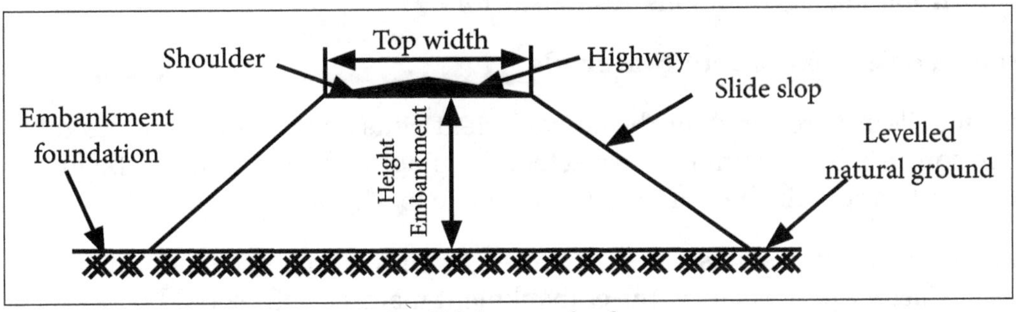

Details of highway embankment.

Embankment Design

Important elements in embankment design are:

- Selection of embankment dimension.
- Stability of slopes.
- Settlement analysis of embankment.
- Selection of embankment materials.
- Drainage of embankment.
- Embankment on weak soil.

Selection of Embankment Dimension

1. Roadway Width

Roadway widths are fixed as per Indian Standard recommendation for single lane and two lane roads.

In case of broader roads with more than two lanes, the road width has to be worked out to suit the number of lanes, medians and shoulders.

2. Height

It is based on the desired grade line of the highway and the ground profile or topography.

The height of the fill is sometimes governed by stability of foundations, particularly when the foundation soil is weak.

3. Side Slopes

In the view of erosion control, a flat slope is to be preferred. It involves, more earthwork but ensures safety of traffic.

2 horizontal and 1 vertical slope in un-inundated conditions and 3 horizontal and 1 vertical un-inundated conditions are satisfactory.

Selection of Embankment Materials

Granular soils are preferred for the embankment. Silts and clays are not desirable and organic soils are to be summarily rejected. Keeping the lead and lift to be low as the main consideration, the best of the soils available locally is often selected.

In case of foundation, soil is weak; light-weight fill material like cinder may be used. For a satisfactory performance, the embankment material soils should possess the following properties:

- A soil with large strength and of high stability. A well-graded soil gives high strength and stability should be preferred. Such a selection of soil will give an economical pavement design.
- Compared to a fine-grained soil, course-grained are preferred.
- Highly plastic soils should be avoided as it may exhibit large changes in volume which may lead to disruption of pavements.
- Soil should have good drainage properties. In general well-graded coarse fractions exhibit good drainage properties and clayey soils should be avoided.
- Soils which could be compacted to high densities should be preferred.
- It is undesirable to use soils which are susceptible for capillary rise.
- Frost-susceptible soils should be avoided.
- Soils with high organic matter and with high presence of sodium sulphate should be avoided.

As per Indian Road Congress recommendations, the general guide to selection of soils on the basis of anticipated embankment performance is given in table.

Table: General guide to selection of soils on the basis of anticipated embankment performance.

PRA Classification	Most Probable I.S. Classification	Possible I.S. Classification	Visual Description	Max. Dry Density gm/c.c	OMC	Anticipated Embankment Performance
A-1	GW, GP, GM, SW,SP,SM	-	Granular materials	1.84 - 2.28	7-15	Good to excellent
A-2	GM,GC, SM,SC	-	Granular materials with soil	1.76 - 2.16	9-18	Fair to excellent

A-3	SP	-	Sand	1.76 -1.84	9-15	Fair to good
A-4	ML, MH, OL,OH	CL, SM, SC	Sandy silts and silts	1.52 - 2.08	10-20	Poor to good
A-5	MH, OH	-	Elastic silts and silts	1.36-1.60	20-35	Unsatisfactory
A-6	CL, CI	MH, OH, SC	Silt clay	1.52-1.92	10-30	Poor to good
A-7	MH, CI, CH.OH	sc	Clay	1.36-1.84	15-35	Poor to fair

Drainage of Embankment

Special considerations have to be exercised for drainage of high embankment and protection of slopes from erosion due to surface flow.

1. Drainage of High Embankment

Embankments with height more than 8 m need consideration. Such embankments have steep longitudinal gradients particularly at approaches to bridges. Serious damages can occur if the rain water is allowed to leave the carriage way in an uncontrolled manner.

It causes erosion of slopes and shoulders. This may undetermined the road pavement.

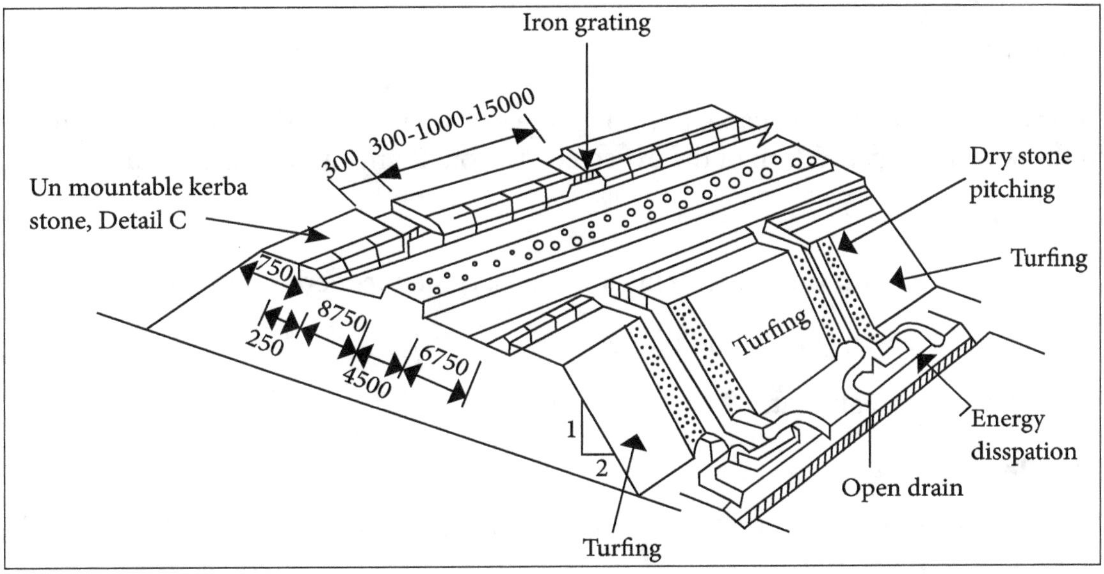

Typical drainage arrangement in high embankment.

It is necessary to provide longitudinal drains from where the water is led down the slopes by means of stepped outfalls or lined chutes. Water from these outfalls finally discharges into the open drain at the bottom. Embankments slopes in between the stepped outfalls or lined chutes need to be protected from surface erosion.

2. Protection of Slope

Run-off from the pavement surface and the shoulders ultimately falls over the slopes of the embankments. Standing flood water against the road embankment in flood-prone zone causes erosion during draw-down. Erosion due to flows may cause the following harmful effects:

- Reduction in the effective width of roadway and shoulders due to the formation of rain cuts.
- Stability of the slope is endangered due to increase in driving force due to saturation of the soil and reduction of shear strength.
- Continued additional expenditure or maintenance.

Different factors contribute for erosion such as soil type, slope of embankment, height of embankment, intensity of rainfall, etc.

The following aspects are to be considered:

- High approaches to over-bridges.
- Embankments in loose sandy soils.
- Heavy rainfall zones.

Treatments

Variety of treatment is available to protect the embankment and cut slopes depending on the severity of protection of the expected erosion. Some of the remedies are:

- Turfing.
- Pitching (with riprap, cement sand slabs, cement concrete slabs or bridges).
- Raised curbs, gutters and flumes.
- Bitumen treatment.

Turfing

It is the most well-known and cheapest form of slope protection. Growth of turf depends on the type of soil.

Turf can be grown by seeding. Seeding is done in flatter slopes whereas seeds are preferred in steeper slopes.

Pitching

This is particularly done for high embankments in flood zones and heavy rainfall areas.

In order to provide outlet for seepage of water in the embankment, open jointed pitching is preferred. A typical pitching is shown in figure.

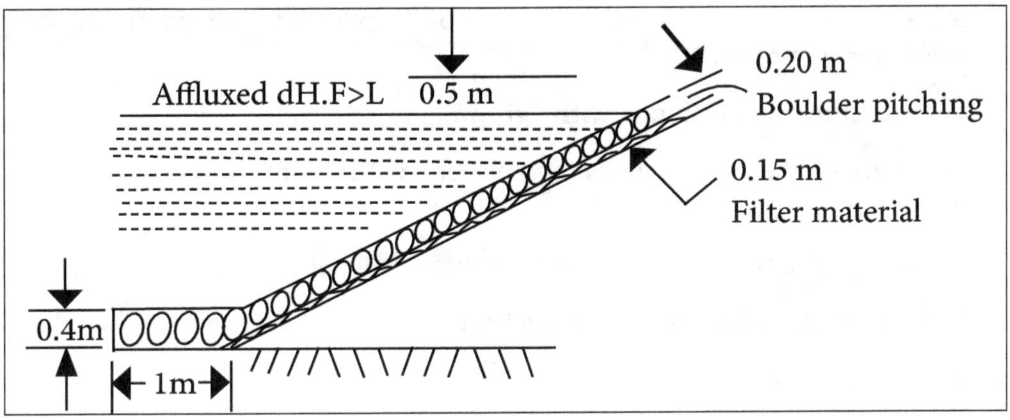

Pitching embankment slope.

Curbs

Curbs, gutters and flumes are provided where high approaches to over-bridges are needed in heavy rainfall zones. Dimensions, opening of flumes and slope depend on the demand.

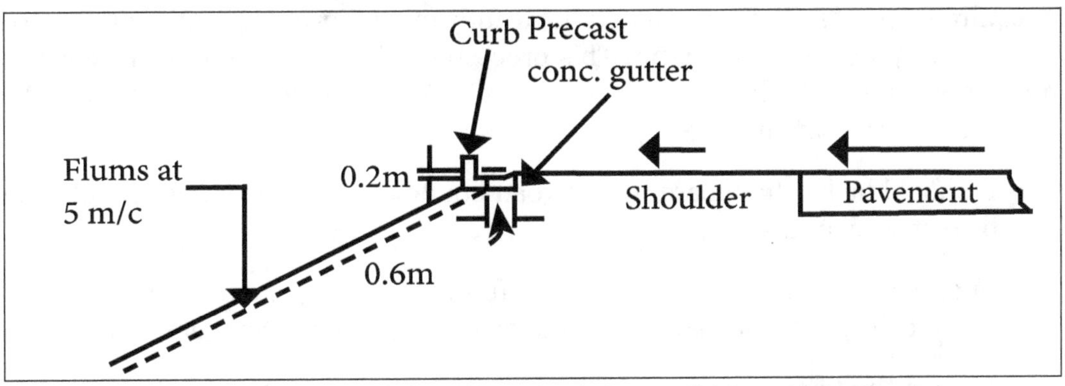

Curbs, gutter and flume arrangement for drainage of highway embankment.

Bituminous Treatment

This treatment is given in different ways. The asphalt is sprayed as an emulsion which helps in a number of ways. It acts down the susceptibility to erosion. It helps to raise the soil temperature by absorbing light rays and promoting the emergence of tiny saplings. The other approach consists of lying and compacting asphalt mixes or spraying a low viscosity cut-back.

Embankments on Weak Soils

In locations on weak soils special design and construction features are adopted. If unsuitable soil conditions are encountered at the site of a proposed structure, one of the following five procedures is adopted:

- Bypassing the unsuitable soil by means of deep foundation extending to a suitable bearing material.
- The structure is redesigned with light weight material.
- By removing the poor material and either treat it to improve and replace it by a suitable material.
- By treating the soil in place to improve its properties.
- By adopting suitable stage construction procedure with the use of sand drains.

Embankment Construction

Construction of embankment may be done either by rolling in relatively thin layers or by hydraulic fills. The former method is called rolled-earth method which is preferred in highway embankments. Embankment construction is performed in different stages as detailed below.

Borrow-pit

The required type of soil may not be available fully in quality and quantity but are to be processed in a pit called borrow-pit. This processed soil is only used for the construction of embankment. Following practice has been recommended by the Indian Road Congress (1961) for borrow- pits:

- It should be located nearer to the road boundary as far as possible and should be rectangular in shape.
- In order to provide due allowance for future widening of embankment, no borrow pit should be located within 5 m from the toe of the final embankment.
- The depth of borrow-pit should be such that they do not cut an imaginary line having a slope of 1 in 4 projected from the edge of the bank. The maximum depth should be limited to 1.5 m.
- Borrow-pits should not be dug continuously. Ridges should not be less than 8 m width and should be left at intervals not exceeding 300 m. In order to facilitate drainage small drains should be cut through the ridges.
- In temporarily acquired land, the depth of borrow pits should not be greater than 45 cm.
- A collection of water in borrow-pit may give room for breeding of mosquitoes; they should not be dug within 0.8 km of towns and villages.

Cleaning of Natural Ground

Clearing and grubbing are the primary operation to be carried out in highway earthwork. It consists of removing trees, stumps, roots, shrubs, etc. from the area to be occupied by the embankment. Top soil is stripped in the areas where the embankment soil is not conducive to the growth of turf. Stripped material is then laid as a blanket on the embankment slopes.

Compaction of Original Ground

The dry density of the original ground should be 90% of the proctor's density. The original ground is brought to the required density by loosening the soil, watering and rolling in layers of 25 cm. This treatment has to be extended to a depth of 0.5 m. If the soil 15 cm beneath this depth does not have a compaction of at least 90%, it should be made up to at least 95%.

Compaction of Embankment

Embankment is constructed in layers by rolling. First the soil is dumped on the natural ground. The water content should be OMC ± 2%.

Required water is added at the borrow-pit or at the site. Soil clods should be broken to have a maximum size of 15 cm. In the top of the embankment, the maximum size should be less than 6 cm.

The soil is spread and depth of each layer shall be generally restricted to 25 cm. Mechanical equipment is needed for achieving the desired compaction.

Type of rollers used is sheep foot roller, pneumatic tyre roller, vibratory roller and power rammers. Compaction of each layer, called lift, should achieve 90 to 95% of proctor's density.

Higher densities are desired in the top 0.5 m of the embankment immediately below the pavement than in the other portions.

Density requirements for embankments and subgrade materials, as per IRC (2001), are given in table.

Table: Density Requirements of Embankment and Subgrade.

S. No.	Type of Work	Maximum Laboratory Dry unit Weight kN/m3.
1.	Embankments up to 3 m height, not subjected to extensive flooding.	Not less than 15.2
2.	Embankments exceeding 3m height or embankments of any height to long periods of inundation.	Not less than 16.0
3.	Subgrade and earthen shoulders verges/backfill.	Not less than 17.5

4

Highway Construction Materials and Practice

4.1 Highway Construction Materials: Properties and Testing Methods

Soil

Properties of subgrade soil are important for the design of pavement structure to withstand adverse weather conditions. Soil is also used in embankment construction which should provide adequate stability and also incompressible. It is an unconsolidated material that has resulted from the disintegration of rocks. It constitutes the earth's surface.

The type and characteristics of soil depend largely on its origin. Soil deposits constitute an assemblage of solid particles with formation of certain pore or void spaces. Voids are in turn filled with a gas or liquid or both. Three components viz., solid, gas and liquid play an appreciable role in the fundamental behaviour of soil.

Desirable Properties of Soils

The properties of a soil depend on size, shape, surface texture and the chemical composition of its grains. Density and moisture content influence to great extent the engineering properties of a soil.

Soils to be used as a highway material should have the following properties:

- Short and long term stability of the subgrade and slopes of embankment.
- Compressibility within permissible limits.
- Adequate permeability.
- Compaction should be ease and economical.
- Minimum volume change at all weather conditions.

Tests on Soils

The suitability of soil is ascertained from various laboratory and field tests. Apart from the fundamental tests to determine the shear strength properties, permeability properties and settlement properties, the following laboratory and field tests are discussed.

1. Field Density Tests

- Core - cutter method.
- Sand replacement method.

2. California Bearing Ratio Test

- Laboratory test.
- Field test.

3. Plate Bearing Test

Field density of a soil is essential to assess the strength of a subgrade and shoulders. California Bearing Ratio test is an Ad-hoc test which gives an empirical strength number which is used to assess the bearing power of a subgrade. Plate bearing test is used to find the modulus of subgrade reaction which is used in the pavement design.

Field Density Tests

Theory

Dry density of a soil in the field is called the field density or in - situ dry density or in - place dry density. Field density is found using adopting core - cutter method (IS: 2720 - Part 29, 1975) or sand - replacement method (IS: 2720 Part 20,1974).

1. Core - Cutter Method

Apparatus Required

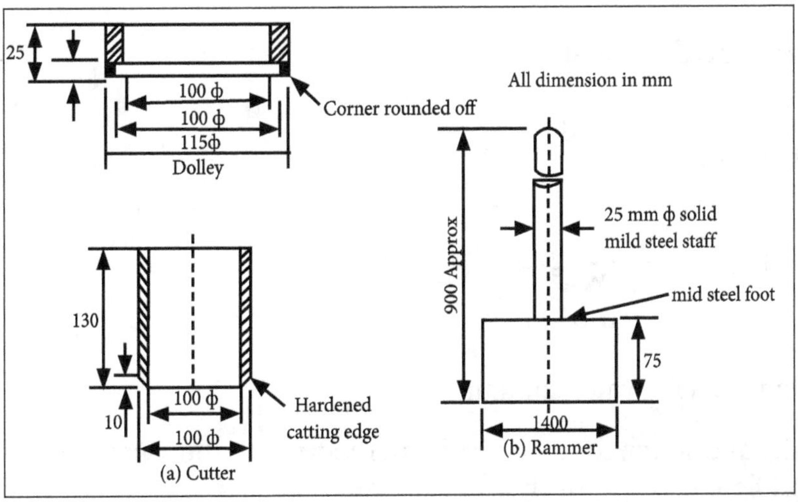

Apparatus for core cutter method.

- Cylindrical core cutter of steel with steel dolly and steel rammer.
- Steel rule, palette knife, straight edge.
- Balance of 1g sensitivity.
- Apparatus for water content determination.
- Apparatus for extracting samples from the cutter.

Test Procedure

- The inner dimensions of the core - cutter is measured (nearest to 0.25 mm) and its volume is found (V).
- The weight of the cutter is found without the dolly.
- A small area of about 300 mm2 is cleared and leveled where the in -place density is to be determined.
- The cutter is placed on the leveled surface. The dolly is positioned on the cutter.
- The cutter is advanced into the subsoil by the help of rammer until about 15 mm of the dolly protrudes above the surface.
- The soil around the cutter is dug, using a spade or pickaxe and the cutter is bodily removed with soil allowing some soil to project from the lower end of the cutter. The top and bottom of the cutter are trimmed by means of a palette knife and straight edge.
- The cutter with soil and without dolly is weighed (M_{sc}).
- The soil is removed from the cutter and the water content determined.

Computation and Result

Then Bulk density, $\rho = \frac{M_{sc} - M_c}{V_c}$ g/cc.

The Field density, $\rho_d = \frac{\rho}{1 + \frac{w}{100}}$ g/cc.

Where,

w is the percentage field water content.

The dry density and water content of soil are reported to the second decimal place and two significant figures respectively.

Applications and Limitations

- The core cutter method is quite useful for fine - grained soils
- It cannot be used on soils with more gravel or non - cohesive soils.
- It is a convenient and fast test.

As per IS norms repeat determinations should be done and average result should be reported.

This method is less accurate compared to sand replacement method.

2. Sand - Replacement Method

Apparatus Required

- Sand pouring cylinder.
- Clean and closely graded natural sand passing 1 mm IS sieve but retained on 600 μm IS sieve is cleaned.
- Cylindrical calibration can.
- Metal tray with hole - 300 mm2, 40 mm deep, with a 100 mm hole in the center.
- Tools for excavating holes.
- Glass plate - 450 mm2, 9 mm thick.
- Balance of 1 g sensitivity.
- Apparatus for water content determination.
- Metal tray or container.

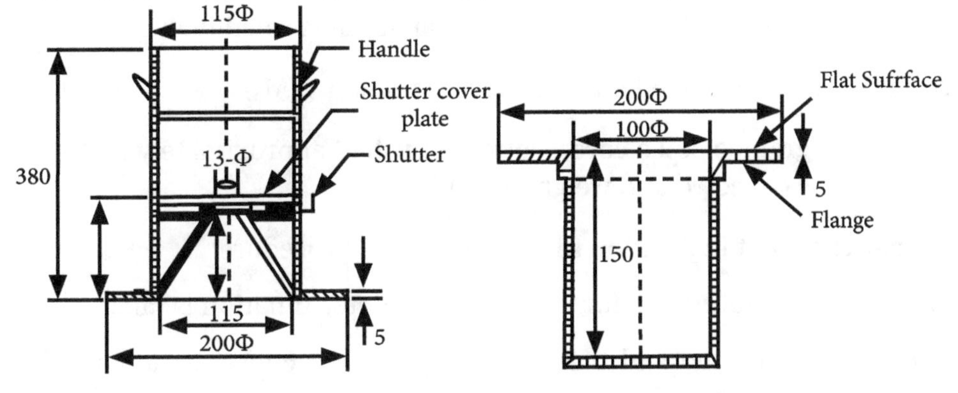

(a) Sand pouriing cylinder (b) Calibration can

Apparatus for sand replacement method.

Test Procedure

- The pouring cylinder is filled with clean sand till the level of sand is about 10 mm from the top and the weight, $M_{1,}$ is found.
- Mass is maintained constant throughout the test for which the pouring cylinder has to be calibrated.
- Pouring cylinder is placed on a glass plate and the top is closed when the conical portion has been filled.
- Sand on the glass plate is carefully collected and the sand is weighed. Steps i) to iii) are repeated atleast three times and the average mass of sand filling the cone is found (M_2).
- Internal dimensions and volume of the calibration can be measured.
- The can is filled with water up to the brim and the mass is found. From the mass the volume is found and checked with the previous one and average volume, V, is determined.
- The pouring cylinder is placed concentrically on top of the calibration can with initial mass, M_1. The shutter is opened and the sand is allowed to fill the can. The cylinder is tapped so as to ensure that the can and the conical portion are completely filled with sand. The cylinder is weighed.
- The step (vi) is repeated thrice and the average mass M_3 of the cylinder after filling the cone and the can is recorded.
- An area of about 450 mm2 is cleaned and levelled.
- The square tray with a central hole is placed on the prepared surface.

A circular hole of 100 mm diameter and 150 mm depth is roughly dug.

All the soil is carefully collected and the mass is found as (M_s).

In case of a fine - grained soil a core - cutter is pushed approximately up to a depth of 100 mm, the soil is collected and weighed as M_s.

The core - cutter is kept in position till the rest of the procedure:

- Some representative soil for water content determination is taken.
- The tray is removed and the pouring cylinder is concentrically placed on the hole with initial mass M_1.

The shutter is opened and the sand is allowed to fill the core and hole. The cylinder is weighed as (M_4).

Computations and Results

Mass of sand filling calibration can = M_a

$$= (M_1 - M_3 - M_2)g$$

Bulk density of sand, $\rho_{sd} = \frac{M_s}{V}$ g/cc.

Mass of sand required to fill the excavated hole, $M_b = (M_1 - M_4 - M_2)$ g.

Bulk density of soil, $\rho = (M_s / M_b) \times \rho_{sd}$ g/cc.

Dry density of soil, $\rho_d = \frac{\rho}{1 + \frac{w}{100}}$ g/cc.

The dry density and water content of the soil are reported to the second decimal place and to two significant figures.

Applications and Limitations

In granular soils with a little or no fines there is a possibility of slumping of the sides of the excavated soil in the hole. In order to avoid bulking of sand due to atmospheric humidity, it is necessary to check the bulk density during each day's work.

For fine and medium - grained soils, a three - litre pouring cylinder is used whereas for fine, medium and coarse - grained soils a greater capacity of 16.5 litres is used. As the in - place density of a soil varies from point to point, tests are conducted at different locations close to each other and the mean value is recommended.

4.2 CBR Test for Subgrade

California Bearing Ratio Test referred as CBR Test is an ad-hoc penetration test developed by the California State Highway Department of the USA. It is used for the evaluation of subgrade strengths for roads and pavements.

It is expressed as the percentage of force per unit area required to penetrate a soil mass with a circular plunger of 50 mm diameter at a rate of 1.25 mm/min compared with that required for the corresponding penetration in a standard material.

Laboratory Method:

1. Apparatus Required

- A mould shall be made of non-corrodable metal and be cylindrical in shape with an inside diameter of 150 mm and of height 175 mm. It shall be provided with a detachable metal extension collar of 50 mm in height and a detachable perforated base plate of 10 mm thick. The perforations in the base plate shall not exceed 1.5 mm in diameter.

- A collar of 50 mm height and 150 mm diameter.
- Circular metal spacer disc of 148 mm diameter and 47.7 mm in height.
- Metal rammer having a weight of 2.6 kg with a drop of 310 mm or having a weight of 4.89 kg with a drop of 450 mm.
- Expansion measuring apparatus consists of a plate 148 mm in diameter with perforations of 1.5 mm in diameter with a threaded stem at the center. A metal tripod fitting over the rim of the mould with a dial gauge clamped for measuring the amount of swell during soaking.
- Annular metal weight and several slotted weights weighing 2.5 kg each.
- Loading machine with a capacity of 50 kN (5000 kg approximately) with a rate of strain of 1.25 mm/min.

2. Preparation of Test Specimens

The test may be performed on:

- Undisturbed specimens or
- Re-moulded specimens which may be compacted either statically or dynamically.

(i) Undisturbed Specimen

This is obtained by fitting a cutting edge of 150 mm diameter to the mould and pushing the mould as gently as possible into the ground. As the mould is pushed in, the soil is dug from the outside. When the mould is full of soil, it is removed by under - digging. The top and bottom surfaces are trimmed flat so as to get a specimen of the required length ready for testing. If the soil is hard and the mould cannot be pressed, then a large undisturbed lump of soil is cut out, from which the required specimen for the mould is made. If the specimen is loose in the mould, the annular cavity is filled with paraffin wax. The density of the soil and the water content are determined so as to determine the dry density.

(ii) Re-moulded Specimen

The dry density of the re-moulded specimen has to be decided which may be the field density, maximum dry density or any other density. Accordingly the moulding water content may be OMC or field moisture content. Soil for the re-moulded specimen shall pass through a 19 mm sieve.

Allowance for large size particles is made by replacing the soil by an equal amount of material passing a 19 mm sieve and retaining in a 4.75 mm sieve. Soil is prepared and compacted in the mould either statically or dynamically as required.

In both cases of compaction if it is intended that the specimen is to be soaked, the water content of the soil before and after compaction is determined. If the specimen is not to be soaked, a representative sample is taken for water content determination.

3. Test Procedure

(i) Test for Swelling

A filter paper is placed over the specimen and the adjustable stem and the perforated plates are placed on the soil specimen. Weights are then applied to produce a surcharge equal to the weight of the base material and pavement expected in actual construction within 2.5 kg. Whole weight should not be less than 5 kg.

Whole mould and weights are immersed in a tank of water allowing free access of water to the top and bottom of the specimen. Tripod for the expansion measuring device shall be mounted on the edge of the mould and the initial dial gauge reading recorded.

Set-up is kept as such without any disturbance for 96 hours. Readings are noted down every day against the time of reading. During this entire period water level in the tank is maintained.

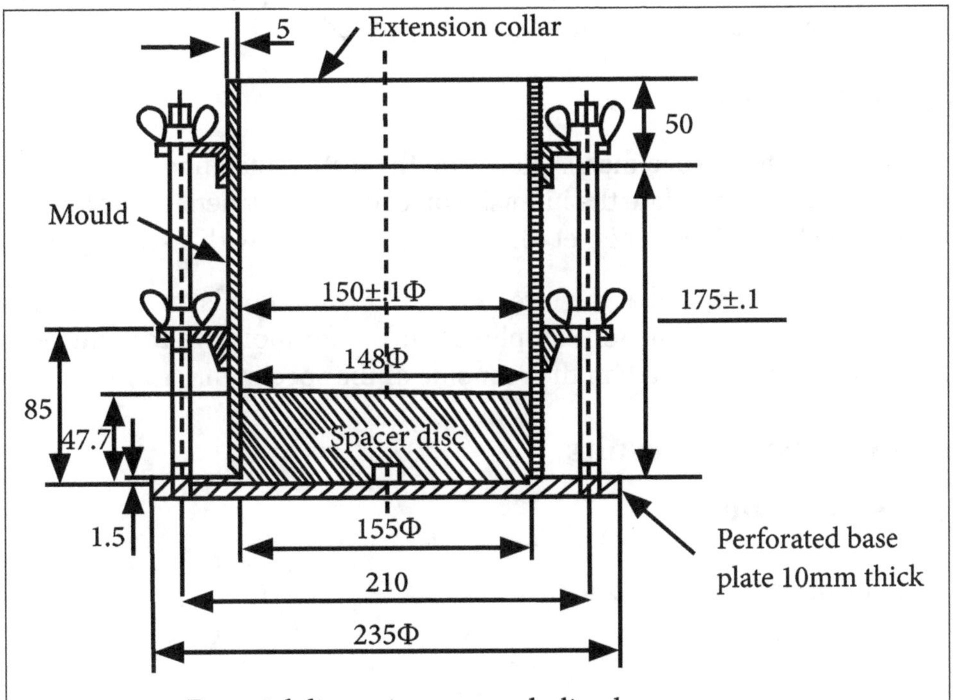

Typical mould with extension collar.

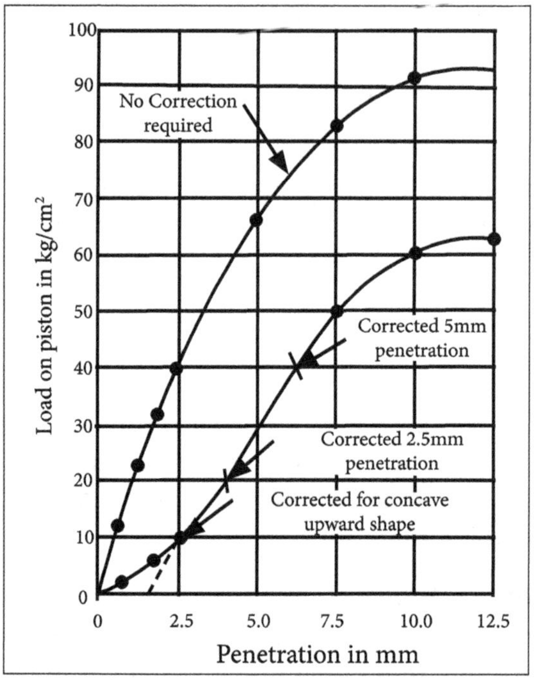

Correction of load penetration curves.

At the end of the soaking period, change in the dial reading is noted. The tripod and the mould are removed from the water tank. The specimen is allowed to drain for 15 minutes downwards. Weights, perforated plate and filter paper are removed. Weight of the mould along with the soaked soil specimen is found.

(ii) Penetration Test

The mould along with the specimen is placed on the testing machine. Required number of surcharge weights to simulate the intensity of loading equivalent to the base material and pavement is placed. Plunger is set to touch the specimen with a seating load of 4 kg.

Load on the plunger is applied at the rate of 1.25 mm/min. Loads are noted at different penetration levels till 12.5 mm. Soil samples are taken for moisture content determination. The test may be conducted on the rear side of the specimen.

4. Computations and Results

(i) Expansion Ratio

$$\text{Expansion ratio} = \frac{d_f - d_s}{h} \times 100$$

Where,

d_f is the final dial gauge reading in mm,

d_s is the initial dial gauge reading in mm and

h is the height of the specimen in mm.

The expansion ratio is used to identify qualitatively the potential expansiveness of soil.

(ii) CBR Value

The load penetration curve is plotted. The curve is usually convex upwards although the initial portion of the curve may be convex downwards due to surface irregulations. A correction shall be made. The CBR values are usually calculated for penetrations of 2.5 and 5 mm.

$$\text{Then, } CBR = \frac{P_T}{P_s} \times 100$$

Where,

P_T is the corrected unit (or total) test load corresponding to the chosen penetration from the load - penetration curve and

P_s the unit (or total) standard load for the soil depth of penetration as for P_T taken from below table.

Table: Standard loads

Penetration Depth mm	Unit standard Load kg/ cm_2	Total Standard Load kg
2.5	70	1.370
5.0	105	2.055
7.5	134	2.630
10.0	162	3.180
12.5	183	3.600

Generally, the CBR value at 2.5 mm penetration will be greater than that at 5.0 mm penetration. In such a case, the former shall be taken as the CBR value for design purposes. If the CBR corresponding to a penetration of 5 mm exceeds that of 2.5 mm, the test shall be repeated.

4.3 Tests on Aggregate and Bitumen

Test for Road Aggregates

Tests for road aggregates to check the desirable properties are to be conducted. The relevant tests are:

- Crushing test,

- Abrasion test,
- Los Angeles test,
- Deval's test,
- Impact test,
- Soundness test.

Crushing test

1. Theory

The aggregate crushing value gives a relative measure of the resistance of an aggregate to crushing under a gradually applied compressive load. The standard aggregate crushing test is made on aggregates passing a 12.5 mm IS sieve and retained on 10 mm IS sieve.

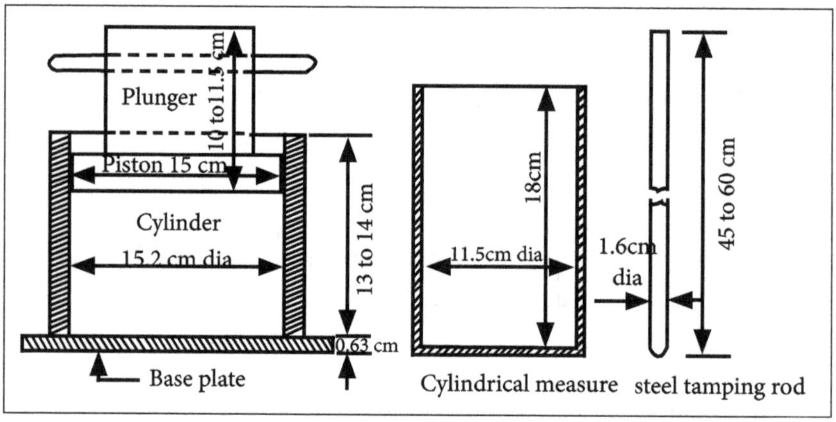

Aggregate crushing test apparatus.

2. Apparatus Required

- A steel cylinder of 15.2 cm diameter and 13 to 14 cm height with a base plate and a plunger.
- Cylindrical measure of 11.5 cm diameter and 18 cm height.
- Tamping rod of 1.4 cm dia and 45 to 50 cm long.
- IS standard sieves.
- Compression testing machine.

3. Test Procedure

- Dry aggregate of 6.5 kg passing 12.5 mm IS sieve and retained on 10 mm IS sieve is taken.

- The aggregate is filled into the standard cylindrical measure in three layers approximately of equal depth.
- Each layer is tamped 25 times with the tamping rod and finally levelled off using the tamping rod as straight edge.
- The weight of the sample contained in the cylindrical measure is measured (M_1). The same weight of the sample is taken for the subsequent test.
- Then the test sample is placed in the test cylinder in three equal layers tamping each layer 25 times.
- Then the plunger is placed on the top of the sample along with the cylinder which is in turn placed on the compression testing machine.
- The sample is loaded uniformly at a rate of 4 tonnes per minute up to a load of 40 tonnes.
- The load is then released and the whole of the material is removed from the cylinder and sieved on 2.36 mm IS sieve.
- The material passed through the sieve is weighed (M_2).

4. Computation and Result

The aggregate crushing value is the percentage of the crushed material passing 2.36 mm IS sieve in terms of the original weight of the specimen.

$$\text{Aggregate crushing value} = \frac{M_2}{M_1} \times 100.$$

5. Applications and Limitations

- Aggregates with low aggregate crushing value are the strong one.
- For a base course the recommended aggregate crushing value should not exceed 45 % and for wearing surface courses should be within 30%.
- For concrete the aggregate crushing value should not be more than 45%.
- If required or if the standard size is not available, other sizes up to 25 mm may be tested. Because of non-homogeneity of aggregates the results will not be comparable with those obtained in the standard test.

Aggregate Abrasion Tests Theory

Aggregates used in the surface course are subjected to wear and tear which is totally referred as abrasion. Abrasion value of aggregates is the difference between the original weight and the final weight of the test specimen tested in a standard abrasion testing machine and represented as a percentage. Indian Standard 2386 (Part IV) of 1963 recommended the use of Los Angeles abrasion testing machine or the Deval's abrasion testing machine.

1. Los - Angeles Abrasion Test

(i) Equipment Required

Los-Angeles abrasion test is based on the principle that the percentage wear is due to the relative rubbing action between the aggregates and steel balls used as abrasive charge. The weight of each charge and the number required depends on the type of grading of test samples.

Table shows the grading of test samples A,B,C, D,E,F and G and Corresponding specified abrasive charge to be used is given in Table. Each charge consists of cast iron spheres or steel spheres approximately 48 mm in diameter and each weighing between 390 to 445 gm.

Table: Gradings of Test Samples.

Sieve Size		Weight in gm Oftest Sample for Grade						
Passing mm	Retained mm	A	B	C	D	E	F	G
80	63	-	-	-	-	2500	-	-
63	50	-	-	-	-	2500	-	-
50	40	-	-	-	-	5000	5000	-
40	25	1250	-	-	-	-	5000	5000
25	20	1250	-	-	-	-	-	5000
20	12.5	1250	2500	-	-	-	-	-
12.5	10	1250	2500	-	-	-	-	-
10	6.3	-	-	2500	-	-	-	-
6.3	4.75	-	-	2500	-	-	-	-
4.75	2.36	-	-	-	5000	-	-	-

Table: Specified Abrasive Charge.

Grading	Number of Spheres	Weight of Charge (gm)
A	12	5000 ± 25
B	11	4584 ± 25
C	8	3330 ± 20
D	6	2500 ± 15
E	12	5000 ± 25
F	12	5000 ± 25
G	12	5000 ± 25

Los Angeles machine - It consists of a hollow cylinder closed at both ends, having inside diameter 70 cm and length 50 cm and mounted so as to rotate about its horizontal axis.

(ii) Test Procedure

- The specified weight of aggregate sample (5 or 10 kg) depending on the gradation is placed in the machine.
- The abrasive charge corresponding to the specified weight of aggregate is added to the machine.
- The machine is rotated at a speed of 20 to 33 rev/min.
- For grading A, B, C and D, the machine is rotated for 500 revolutions.
- For grading E, F and G it is rotated for 1000 revolutions.
- After the completion of the above number of revolutions the material is discharged from the machine.
- Preliminary separation of the sample is made on a sieve coarser then 1.7 mm sieve. Finer portion is than sieved on a 1.7 mm IS sieve.
- The material coarser than 1.7 mm IS sieve is washed, dried in an oven at 1050 to 110°C to a substantially constant weight and accurately weighed to the nearest gram.

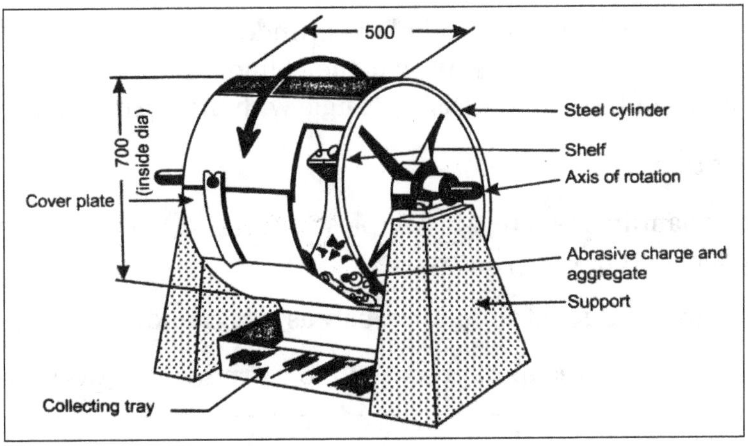

Los Angeles abrasion testing machine.

(iii) Computation and Result

The difference between the original weight (M_1) and the final weight of the test sample (M_2) is expressed as a percentage weight of the test sample.

$$\text{Los}-\text{Angeles Abrasion Value}=\frac{M_1-M_2}{M_2}\times 100.$$

(iv) Applications and Limitations

- The percentage of wear should not be more than 16 % for concrete aggregate.
- For bituminous mixes Los-Angeles abrasion values up to 30% is allowed.
- For base courses like bitumen bound macadam Los Angeles abrasion value up to 50 % is allowed.
- This test is more dependable as rubbing and pounding action taking place inside the machine almost simulates the field condition.
- Further correlation of Los Angeles abrasion value with field performance is already established.

2. Deval Abrasion Test

(i) Equipment Required

- Its principle is almost same as that of Los Angeles abrasion test.
- Here the samples of aggregates are allowed to tumble over in a rattler in the presence of abrasive charge. The abrasive value is expressed as in the Los Angeles abrasion test.
- Here the abrasive charges consist of 6 cast iron or steel sphere of about 4.8 cm diameter and total weight is 2500 g.
- Deval machine consists of two hollow cylinders of 20 cm diameter and 34 cm length mounted in such a way that the cylinders rotate about the horizontal axis. The axis of the cylinders is 30° angle with the horizontal.

(ii) Test Procedure

- Specified quantity of aggregate samples (4 to 5.5 kg) of any one of the specified grading is placed in the cylinder.
- Two tests are conducted simultaneously as two cylinders are available.
- The machine is rotated at a speed of 30 to 33 rpm for 10,000 revolutions.

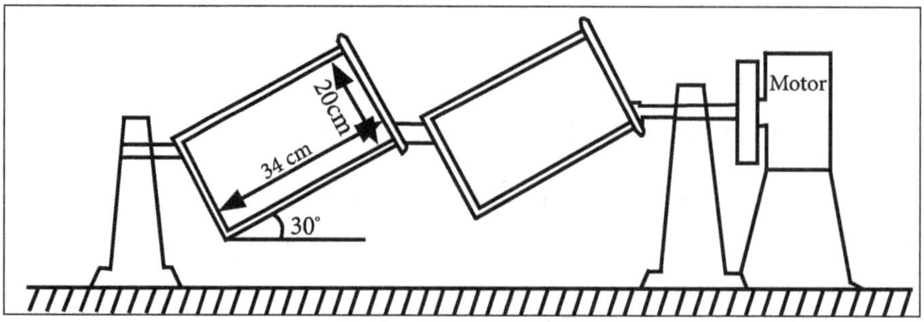

Deval abrasion testing machine.

- After completion of revolutions the material is sieved through 1.7 mm IS sieve and washed, dried and weighed.

(iii) Computation and Result

$$\text{Deval Abrasion Value} = \frac{M_1 - M_2}{M_1} \times 100.$$

Where,

M_1 is the initial weight of the sample.

M_2 is the final weight of the sample.

As two samples are tested at the same time better average value can be obtained. When the test is carried out without charges the test is known as Deval attrition test.

Aggregate Impact Test

1. Theory

It is conducted to evaluate the toughness of stones or resistance of stones to breaking under repeated impact loads. It may be carried out either as a cylindrical stone specimen or stone aggregates. The Page Impact test is one which has been designed to evaluate the toughness of stone specimens.

The impact test conducted on aggregate is known as Aggregate Impact Test which is carried out to evaluate the resistance to impact of aggregates.

The aggregate impact value is different from the crushing test in which the test value indicates a relative measure of resistance of aggregate to shock or impact rather than a gradually applied compressive load.

2. Equipment Required

- Aggregate Impact Testing Machine:
    - It consists of a metal base and a cylindrical cup of 10.2 cm internal diameter and 5 cm in depth. The aggregate sample is placed in this cup. A metal hammer of 13.5 to 14 kg having a free fall from a 38 cm height is arranged to drop through vertical guides.
- Cylindrical Measure.

3. Test Procedure

- The test sample consists of aggregate passing through 12.5 mm and retained on 10 mm IS sieve.
- The aggregate is dried in an oven for a period of four hours at a temperature of 100° to 110° C.

- The sample is filled in the cylindrical measure in 3 layers by tamping each layer by 25 blows.
- The net weight of the aggregate in the cylindrical measure is determined (M_1). This weight of aggregate has to be used for duplicate test on the same material.
- The whole sample is transferred to the cup of the aggregate impact machine and compacted by tamping 25 times.
- The hammer is raised to a height of 380 mm above the upper surface of the cup and allowed to fall freely on the aggregate sample in the cup.
- The test sample shall be subjected to a total 15 such blows each being delivered at an interval of not less than one second.
- The crushed aggregate is recovered from the cup and the whole of it is sieved on 2.36 mm IS sieve. The fraction passing the sieve is weighed to an accuracy of 0.1 gm (M_2).
- The fraction retained on the sieve is also weighed (M_3).
- If the total weight ($M_2 + M_3$) is less than the initial weight M_1 by more than one gm the result shall be discarded and a fresh test is made. Two tests are made.

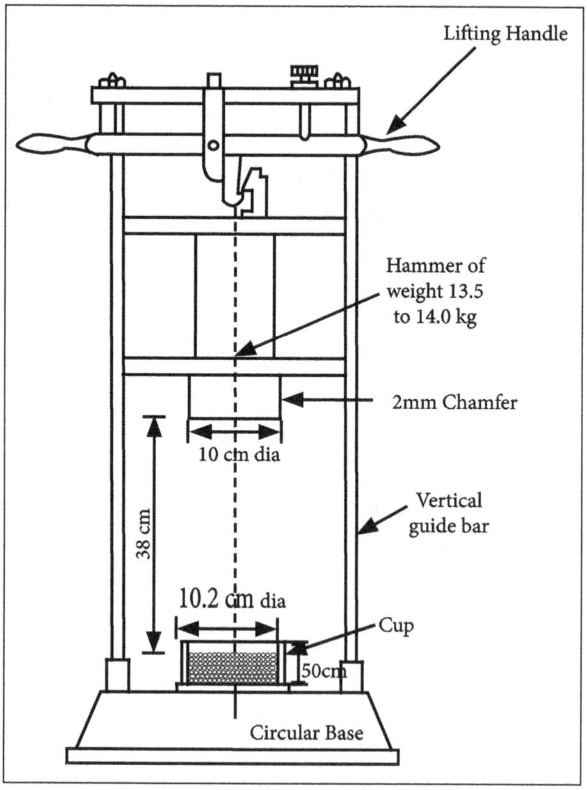

Aggregate impact testing machine.

4. Computation and Result

The ratio of the weight of fines formed to the total sample weight in each test is expressed as percentage. Therefore,

$$\text{Aggregate Impact Value} = \frac{M_2}{M_1} \times 100.$$

5. Applications and Limitations

- The aggregate impact value should not be more than 45 % for aggregates used for concrete.
- Concrete wearing surfaces should use aggregate impact value around 30%.
- Classifications of stones are made on the basis of toughness property considering the aggregate impact values.
- Aggregate impact values less than 10 % is considered as exceptionally strong 10 to 20 % strong and 20 to 30 % satisfactory for road surfacing.

Aggregate Soundness Test

1. Theory

Soundness refers to the ability of aggregate to resist excessive changes in volume as a result of change in physical conditions. The physical conditions that affect the soundness are as follows:

- Freezing and thawing.
- Variation in temperature.
- Alternate wetting and drying under normal conditions.
- Alternate wetting and drying under salt water.

2. Apparatus Required

- Sodium or magnesium sulphate.
- Weighing machine.

3. Test Procedure

- Carefully graded aggregate of known weight (M_1) is taken.
- The sample is soaked in a solution of magnesium sulphate for 16 to 18 hours.
- The sample is dried in an oven at 105 to 110°C to a constant weight. This completes, one cycle.

- Steps (iii) and (iv) are repeated. After the final cycle the sample is examined visually for the evidence of excessive splitting, crumpling or disintegration of grains.
- Sieve analysis is carried out to note the variation in gradation from the original.

4. Computation and Result

The average loss in weight is found which represents the soundness of the aggregate.

The average loss in weight should not exceed 12 or 18 % when tested with sodium sulphate or magnesium sulphate respectively.

Bituminous Materials

Tests on Bituminous Materials

Bitumen is available in a variety of types and grades. To assess the desirability condition different tests are discussed in the subsequent sections. Different tests which are generally used are:

- Penetration test,
- Ductility test,
- Viscosity test,
- Softening point test.

For classification of bitumen and for the study of performance of bituminous pavements, penetration and ductility tests are needed. Softening point, flash and fire point tests are essential to guide the paving technologists on the field. Each test has the limitations as bitumen from different sources behaves differently.

For example bitumen from different sources may have same penetration at a particular temperature but may exhibit different viscous characteristics. For a particular test, value can be true for the given sample only which cannot be correlated with any other fundamental property.

1. Penetration Test

(i) Theory

Hardness or softness of bitumen is assessed by penetration of a standard needle.

It is obtained by measuring the depth in tenth of a millimeter to which a standard loaded needle shall penetrate vertically in 5 seconds.

The bitumen grade is specified in terms of penetration value.

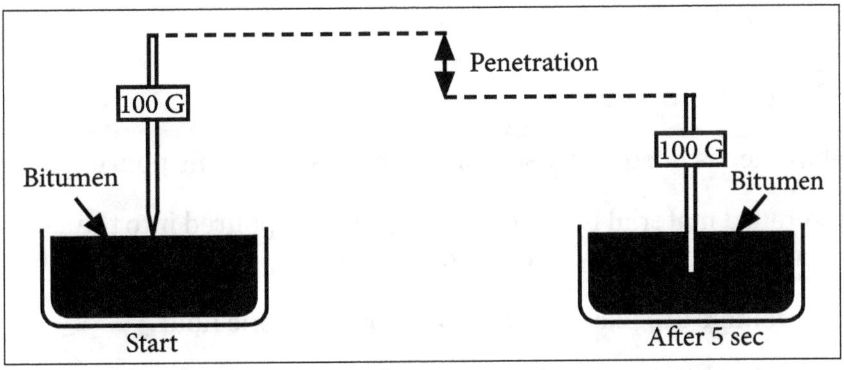

Penetration test concept.

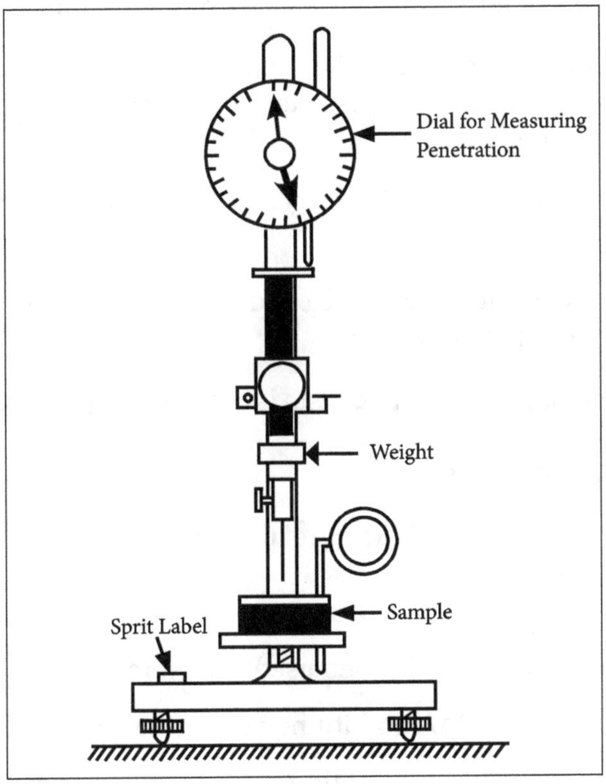

Penetrometer.

(ii) Apparatus Required

- Penetrometer consists of a needle with a device to release and lock a 100 g weight.
- A graduated dial is provided to read the penetration values up to an accuracy of 1/10th of a millimetre.
- A sample container.
- Temperature controlled water bath maintained at 25°C.

- Benzene liquid.

(iii) Test Procedure

- The bitumen is adequately softened such that it attains pouring consistency.
- The softened material is stirred thoroughly and poured into the sample containers to a depth at least 15 mm in excess of the expected penetration.
- The containers are placed in the water bath for one hour.
- The sample container is taken out of the bath and kept on the machine and the needle is made to contact with the surface of the sample.
- The dial is set to zero or the initial reading is taken and the needle is released for 5 seconds. The final reading is taken. The depth of penetration is reported in one - tenth millimetre units.
- After each test the needle is disengaged and the needle point is wiped with benzene and dried.
- At least three penetration tests are conducted on the same sample by testing at distances of at least 10 mm apart.
- The average value of these three tests is reported as penetration value.

(iv) Computation and Result

The bitumen grade is reported with the range of penetration value, e.g., 80 - 100 or 80 /100.

(v) Applications and Limitations

- The penetration values of bitumen used in general varies from 20 to 225.
- The grades 30/40 and 80/100 bitumen are more commonly used.
- In hot weather condition penetration grade bitumen like 30/40 is preferred.
- The penetration test is conducted only for bitumen.
- The penetration value is very much affected by pouring temperature, size of needle, weight placed on the needle and test temperature.

2. Ductility Test

(i) Theory

Ductility is the property of bitumen which permits the material to undergo great deformation without breaking. It is expressed as the distance in centimeters to which a standard briquette of bitumen of 1 sq. cm can be stretched without breaking when the rate of pull is 5 cm per minute at the test temperature of 25°C to 27°C.

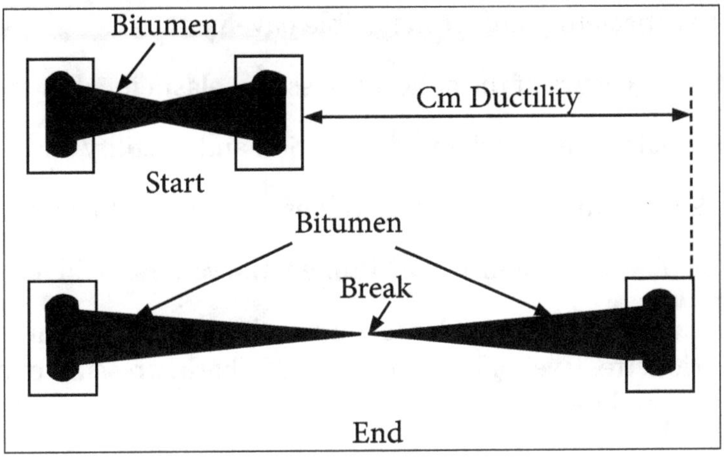

Ductility test

(ii) Apparatus Required

- Ductility test apparatus consists of two briquette mould which can be moved at a pre - calibrated rate.
- Constant temperature water bath.

(iii) Test Procedure

- The bitumen sample is made fluid by heating, poured into two - mould assembly and placed on a brass plate.
- At first the sample and mould assembly is cooled in air and then in water bath maintained at 27°C.
- The excess bitumen material is removed and levelled using a hot knife.
- Then the mould assembly and the sample is replaced in water bath for 85 to 95 minutes.
- The sides of the mould are removed; the clips are hooked on to the machine. The pointer is adjusted to zero.
- One part of the assembly is allowed to pull at the rate of 5 cm per minute.
- The distance up to the point of breaking of thread is noted and the distance is reported in centimetres.

(iv) Computation and Result

The distance from start till the breaking of bitumen is reported in centimetres.

(v) Applications and Limitations

- Ductility test is conducted to find the binding property of a bitumen which serves as a satisfactory binder in flexible pavement.
- This test is a measure of the adhesiveness and elasticity of the bitumen.
- Bitumen should satisfy both the penetration and ductility for a good field work.
- For different bitumen, the ductility values vary from 5 to over 100.
- For satisfactory performance of a bitumen, the value should not be less than 50.
- Ductility values get affected by factors such as pouring temperature, dimensions of briquette, level of briquette in water bath, pressure of air pockets, test temperature and rate of pulling.

3. Viscosity Test

(i) Theory

It is a measure of resistance to flow. Higher the viscosity of a liquid signifies that it can be nearer to semi-solid state of consistency.

It is a function of temperature. It decreases with increase in temperature. Here the time taken by a sample is measured for 50 ml of the material to flow through a standard orifice under standard falling head conditions at a standard temperature.

(ii) Apparatus Required

- Orifice type viscometer,
- Cup,
- Timer.

(iii) Test Procedure

- The cutback bitumen in viscous form is placed in the viscometer.
- The timer is started and flow of the liquid allowed which falls drop by drop in the standard 50 cc spherical glass bulb.
- The time taken in seconds to fill the specific mark in the bulb is noted and expressed as the viscosity.

(iv) Applications and Limitations

- At low viscosity the bituminous binder simply lubricates the aggregate particles instead of forming a uniform film for binding action.

- High viscosity resists the compaction effort.
- Orifice type viscometer indirectly finds the viscosity.
- Furol viscosity is a specific test conducted in liquid bituminous materials.
- Ten millimetre orifice viscometer is recommended for road tar.

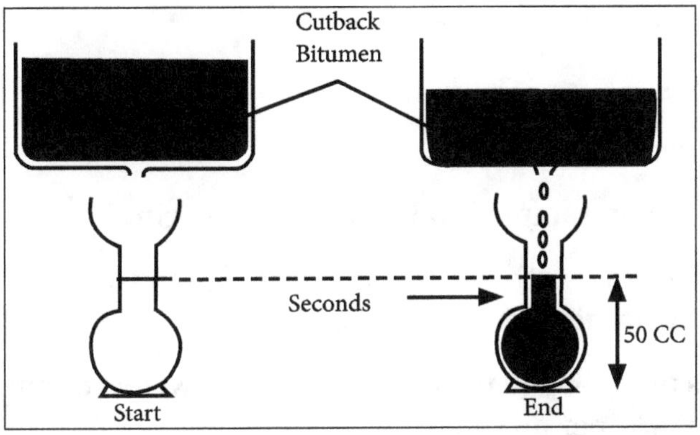

Viscosity test set-up.

- Viscosity is a function of temperature; it decreases with increase in temperature.

4. Softening Point Test

(i) Theory

Softening point is the temperature at which the substance attains a particular degree of softening under specified condition of test.

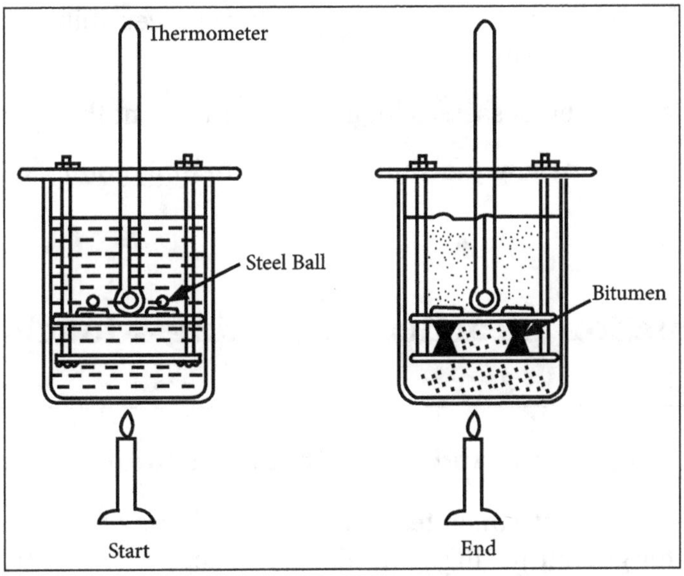

Softening point test set-up.

(ii) Apparatus Required

- Ring and ball set up,
- Glycerine,
- Beaker,
- Burner,
- Thermometer.

(iii) Test Procedure

- The given bitumen sample is taken in the brass ring.
- The brass ring along with the bitumen sample is suspended in water (or glycerine) at a particular temperature.
- A steel ball is placed upon the bitumen and the water (or glycerine) is then heated at a rate of 5°C per minute.
- The bitumen will start melting and start flowing down.
- At a particular temperature, the softened bitumen touches the metal placed at a specified distance below the ring is recorded.

(iv) Computation and Result

The temperature noted is the softening point of the bitumen.

(v) Applications and Limitations

- Higher softening indicates lower temperature susceptibility and is preferred in areas where warm weather exists.
- Hard grade bitumen possesses higher softening point than soft grade bitumen.
- Bitumen used in paving job should have softening point varying between 35° to 70°C.

4.4 Construction Practice Including Modern Materials and Methods

Prefabricated Components and Specialty of Portland Cement Concretes

New generations of specialty concretes have improved one or more aspects of performance and allow for greater flexibility in highway design and construction. High-performance concrete typically has compressive strengths of at least 10,000 psi.

Today, ultra-high-performance concretes with formulations that include silica fume, quartz flour, water reducers and steel or organic fibers have even greater durability and compressive strengths up to 30,000 psi. These enable construction with thinner sections and longer spans.

Latex-modified concrete overlays have been used for many years to extend the life of existing, deteriorating concrete bridge decks by the Virginia DOT, which pioneered the use of very early strength latex-modified concretes for this application.

In high-traffic situations, the added costs of the concrete have been more than offset by savings in traffic-control costs and fewer delays for drivers (Sprinkel, 2006). When the air temperature dips below 400F, costly insulation techniques must be used when pouring concrete for highway projects.

By using commercially available admixtures that depress the freezing point of water, the U.S. Cold-Weather Research and Engineering Laboratory have developed new concrete formulations that retain their strength and durability at temperatures as low as 23°F.

Compared to insulation techniques, this innovation has significantly decreased construction costs and extended the construction season in cold weather regions (Korhonen, 2004). As useful as these and other specialty concretes are nanotechnology and nano engineering techniques, which are still in their infancy have the potential to make even more dramatic improvements in the performance and cost of concrete.

The Superpave Design System

Due to premature failures of hot-mix asphalt pavements in the early 1980s, a well-funded congressionally mandated crash research program was conducted. Its aim is to improve our understanding of asphalt pavements and their performance.

The seven-year Strategic Highway Research Program (SHRP) was managed by the National Research Council. It developed a new system of standard specifications, test methods and engineering practices for the selection of materials and the mix proportions for hot-mix asphalt pavement.

The new system has improved matches between combinations of asphalt binder and crushed stone and the climatic and traffic conditions on specific highways. State departments of transportation (DOTs) spend more than $10 billion annually on these pavements, so even modest improvements in pavement durability and useful life can lead to substantial cost savings for agencies and time savings for motorists (TRB, 2001).

SHRP rolled out the Superpave system in 1993. It took years for individual states and their paving contractors to switch to the new system. It represents a significant departure not only in design but also in the procedures and equipment used for testing.

Each state DOT had to be convinced that the benefits would outweigh the modest additional costs of Superpave mixes, as well as the time and effort to train its staff and acquire necessary equipment.

Bituminous Roads

Bituminous roads are constructed with varying aggregate sizes and compositions. This type of construction is adopted as a wearing course or a surface course. In heavy traffic roads bituminous construction are also adopted for base course.

Compared to cement concrete surfacing, bituminous surfacing takes very less time to use after construction. Black top construction is in large use in developing countries like India where cement is in great demand as a construction material. Further stage construction can be done effectively.

Types of Bituminous Construction

There are various types and methods of construction of bituminous pavements. The following construction techniques are in use:

- Surface dressing or painting.
- Interface treatments.
- Premix materials like:
    - Bituminous bound macadam.
    - Carpet.
    - Bituminous concrete.
    - Sheet asphalt or rolled asphalt.
    - Mastic asphalt.

Bituminous binders adopted in the above methods are straight run bitumen, road tar, cutback or emulsion. The choice of type of binder is based on the type of construction, availability of material, climatic conditions, equipment, etc.

For example, bitumen and tar require heating to bring them to a proper viscosity for their use. Use of this type is called hot-mix technique. On the other hand cutback and emulsion do not require heating as that of bitumen or tar. Use of this material is called cold-mix technique. Based on the technique of mixing and construction, the bituminous constructions are classified as,

- Road-mix methods.
- Central-plant mix methods.

Premix materials construction depends on the gradation. They are classified as open-graded and dense-graded.

Bitumen construction methods for base course are grouted or penetration macadam (either fully-grouted or semi-grouted) and bitumen bound macadam. Different bituminous constructions employed in surface course of pavements are surface dressing, grouted or penetration macadam and seal coat and bituminous premix surfacing.

Surface Dressing

1. Functions and Applications

Main functions of the surface dressing are:

- To provide a dust free pavement surface over the base course.
- To provide full stability to the base course and pavement structure as a whole.
- To prevent infiltration of surface water and thereby form as a water proof layer.

In water-bound macadam roads the binding material, soil slurry, is removed easily by fast moving vehicles. They expose the stone pieces and also make them loose. In such situations a surface dressing by bitumen for a thickness of 2 to 3 cm improves the surface performance better and adds to the overall stability of the structure.

2. Materials Requirements

The bitumen binder may be bitumen or tar or cutback. For a WBM road or renewal of black top surface, the weight required for 10 sq.m. of area are given in table.

Table: Binder requirement.

Base Course Type	Binder Requirement in kg per 10 sq. m. area					
	Single or First Coat			Second Coat		
	Bitumen	Tar	Cutback	Bitumen	Tar	Cutback
WBM	17 to 19.5	17 to 22	19 to 22	10 to 12	12 to 15	12 to 15
Renewal of black to surface	10 to 12	12 to 17	10 to 12	10 to 12	12 to 15	12 to 15

3. Construction Procedure

Surface dressing can be done on an existing road surface as a black top surface or as a new WBM road. In the case of new WBM road surface, the metals are properly rolled after sprinkling water and then the dressing is applied. In case of an old road or existing WBM, the surface is reconditioned to have the design camber and defects rectified.

The surface is cleaned with wire brush such that the aggregates are exposed atleast 1.25

cm. Surface dressing is done in one or two coats depending on the necessity. Over the prepared surface a uniform layer of bituminous binder is applied as given in Table.

In general 2 kg of binder per square meter surface is sufficient. In order to avoid bleeding no excess binder should be used. After spraying, binder chippings of 12 mm thick (for first coat) of 12 mm size of 0.14 to 0.15 cm of materials for 10 sq.m. area is used. For second coat 9 mm thick of 9 mm size of 0.09 to 0.11 cm of material for 10 sq.m. area is applied.

Rolling is started after this from the edges towards the center longitudinally with an overlapping not less than half of the width of rolling wheel. After completion of half of the surface the rolling is started from the other edge. The rolling is continued till all the particles are properly interlocked. This forms the final rolling in case of single coat.

In the case of additional coat, the binder is applied to the prepared surface and chippings spread as per requirement and rolled finally. After ensuring proper cross sections and 24 hours after rolling the road may be opened to traffic.

Interface Treatment

1. Functions and Applications

Depending on the condition of the existing base the following interface treatments are recommended:

- Prime coat is the first coat of bituminous surfacing done on an existing previous texture base.
- Tack coat is the bituminous surfacing done over an existing cement concrete top or already existing black top surface road.
- Seal coat is a final thin coat over certain previous bitumen pavements.

The functions of a seal coat are:

- To increase the skin resistance of the surface.
- To make the surface water proof.
- To increase the life of the surface.

Penetration of grouted machadam is the process by which the bitumen is spread over the compacted aggregate surface to allow the bitumen to fill voids and binding the aggregates together. When the bitumen penetrates fully it is called full grout.

Generally the above treatments are given to existing WBM road surfaces. It is essential that adequate bond should exist between the newly laid bituminous layer and the existing surface at the interface.

2. Construction Procedure

The existing surface is lightly brushed and reconditioned. Priming coat is applied if required.

Additional course aggregates are spread with proper protection of edges. The profile of the cross section is checked with template. A 10 ton roller is used to roll the aggregates when dry.

Rolling is continued till crushing of aggregates is ensured. The profile of the dry-rolled surface is checked for its correctness. Over the dry and compacted surface, the specific binder (bitumen) is uniformly spread either with pressure distributor or hand sprayer.

For a compacted surface of one sq.m. with a thickness of 5 cm and 7.5 cm, 50 kg and 68 kg bitumen respectively are required. After this process, chippings or key aggregates are spread and rolled. Key aggregates of 10 to 12 mm size are used. Volume of material needed varies from 0.15 to 0.18 m3 respectively.

Over this surface premixed sand bitumen or surface dress type seal coat is applied after a few days or immediately.

Premixed Method - Bitumen Bound Macadam

1. Functions

Bitumen Bound Macadam is used in the following areas:

- In hilly regions e they resist frost action better than normal base courses.
- In regions of high water table, they act as impervious cut off.
- By using them, superior quality aggregates may be replaced by medium quality aggregates.

The binder and aggregates are mixed prior to placement and the finished thickness is kept from 5 to 7.5 cm. Size of aggregate varies with the thickness. For a 7.5 cm thick, the maximum size is limited to 37 mm.

2. Construction Procedure

The existing surface is reconditioned to proper cross section and the surface is cleaned. On the prepared and cleaned surface, a thin layer of binder (prime or teak coat) is applied on a 10 sq.m surface and 4.0 to 7.5 kg of binder is to be used for black top surface or 7.5 to 10 kg for untreated WBM surfaces.

Aggregates of proper gradation and binder are separately heated at about 120°C and then mixed in a mixture. The mixture is placed on the already prepared surface and uniformly spread for the required thickness with rakes. The cross section is again checked.

Rolling is done as early as possible after placing premixed material with a 8 to 10 tonne roller. The rolling is started from the edges and processed towards the center with uniform overlapping. Wheels of the roller should be kept wet while rolling so as to avoid sticking of mixed material on the wheels.

Premixed Method-Bituminous Carpet

1. Materials

Bituminous carpet is a premix prepared from 10 to 12 mm size chippings with sand and bitumen binder. Thickness varies from 2 to 2.5 cm. It is used for surface coarse layer. Sand-bitumen carpet mix containing no coarse aggregate is used as a wearing course.

2. Construction Procedure

The undulations and the pot-holes, if any on the existing road are reconditioned properly with pre-coated aggregates. In case of old bitumen or WBM road, a tack coat is applied on the surface. If the surface is made out of soft aggregate, a prime coat is applied.

Bitumen binder is heated to the specified temperature and tack or prime coat is applied. Chippings and bitumen are heated separately to the specified temperature and then mixed so as to form a homogeneous mix.

The mix is spread over the prepared surface with suitable rakes. The cross section is once again checked before rolling. At a stretch, the pre-mixed material is laid for a length of 15 meters and rolled.

Wheel surfaces are kept wet with the help of wet gunny bags or sprinkling water to avoid adhesion of mix on the wheels. Rolling is continued till it stopped showing wave in front of the roller. In low rainfall areas; premixed medium coarse sand is applied on the carpet.

For a 10 sq.m. area, 6.83 kg of binder and 0.06 m3 of sand are required. In high rainfall areas, a liquid seal sprayed with a layer of chippings applied over the carpet. Chippings of 6 mm size of 0.091 m3 per 10 sq.m area are used. Then the chippings are rolled with a light roller. The pavement may be allowed for traffic after 24 hours of the final rolling.

Premixed Method-Bituminous Concrete

1. Functions and Materials

Bituminous concrete surfacing has the following functions:

- It provides a resilient, dust-free water proof, smooth road surface.
- It resists heavy and mixed vehicles in a better way.

- It has high durability and riding qualities as that of cement concrete pavements.

In these method coarse and fine aggregates along with bitumen binder is used to give a dense mix. Thickness depends on the traffic intensity and on the quality of base course. Aggregates and binders are separately heated and then mixed till all the aggregates are coated with bitumen. Mastic asphalt is also used as a binder.

Bituminous concrete surfacing is done in two layers:

- Binder course,
- Wearing course.

The coarse to fine aggregate ratio is 2:1. Binder used is 7.5 % of the mix by weight.

2. Construction Procedure

The existing base course is reconditioned at least one week before laying the binder course. Then the bitumen course layer will be laid. This is also a hot-mix process.

The hot mix is collected and spread over the prepared surface. The camber and thickness of bitumen layer is checked. The placed concrete is rolled by a roller at a speed not more than 5 km/hour. The number of passes required to attain the final desired thickness depends on the thickness.

The initial rolling is done using a 8 to 12 tonne roller followed by fixed wheel pneumatic roller of 15 to 30 tonnes. The wheels are kept damp with water or wet gunny bags.

Cement Concrete Roads

Portland cement concrete pavements are formed with a relatively rich mixture of Portland cement, sand and coarse aggregate laid as a single course. Cement concrete roads are preferred by every user because of excellent riding surface and pleasing appearance.

Construction of cement concrete pavement may be dealt in two parts, viz., construction of concrete pavement slabs and construction of joints.

Necessity of Base

A cement concrete pavement is constructed with or without a base. If a sufficiently strong subgrade is available the concrete slab can be directly constructed on the subgrade, otherwise a base course has to be provided.

Apart from soil subgrade, other factors which are considered for providing a base is Design load, Traffic volume and Economic considerations.

Various purpose of providing base beneath the cement concrete pavement is:

- To perform as a strong supporting layer.
- To prevent the damage due to mud pumping by acting as a capillary cut off.
- To reduce the thickness requirements of slab.

Materials

Materials used in concrete for highway construction are coarse aggregate, fine aggregate (sand), cement and water. In order to modify some of its properties certain additives and admixtures are added.

1. Coarse Aggregate

It consists of natural, crushed stone, gravel or air cooled iron blast furnace slag. Very angular, flaky, elongated or splintery aggregates give a concrete of low workability which should be avoided. To make a quality concrete, crushed aggregates of poor shape should also be avoided. To get a quality concrete, the desirable limits of important properties of coarse aggregate are:

- Aggregate crushing value= 30 % max.
- Aggregate impact value= 30 % max.
- Los Angeles abrasion value= 16 % max.
- Soundness:
    - Average loss in weight after 10 cycles = 12 % max in sodium sulphate 18 % max in magnesium sulphate.
- Gradation in two separate size ranges*
    - 5 to 2.5 cm~ 50 % & 2.5 to 4.75 mm~ 50 %.
    - 40 to 20 mm~ 50 % & 20 to 4.75 mm~ 50 %.

(*Proper proportioning to get the minimum voids).

2. Fine Aggregate

This may be natural sand, crushed stone sands or a combination. When volumetric batching is done, bulking of sand due to the presence of moisture should be accounted. Fine aggregates should not have more than 45 % retained between any two consecutive sieves.

Fineness modulus should not be less than 2.3 and not more than 3.1. It should be free from organic impurities.

3. Cement

Ordinary Portland cement is generally used. The cement should comply with standard specifications prevalent in the country. High-alumina cement may be used in making speedy repairs to roads. Rapid- hardening cement may be used when it is necessary to open traffic with the minimum delay.

In regions of alternate freezing and thawing weather air-entering cement may be used. Irrespective of the type of cement used, the cement should be tested in advance so that 7 and 28 days tests can be completed before the cement is used.

4. Water

It should be clean and clear. Should be free from salt, oil, acid, injurious alkali and vegetable matter or other substances harmful to the finished concrete.

5. Concrete

It has to be proportioned to obtain a minimum modulus of rupture of 35 kg/cm2 on field specimens after 28 days curing or to develop a minimum compressive strength of 280 kg/cm2 at 28 days or higher values as per design.

Modes of Construction

There are two methods of construction of cement concrete slabs:

- Alternate bay method.
- Continuous bay method.

1. Alternate Bay Method

This method involves constructing a bay or one slab in alternate succession leaving the next or intermediate bay. The next construction is done after time gap of one week or so.

For example, the alternate bays X, Y and Z (Figure) are constructed at one stretch. Others, viz., X', Y' and Z' are constructed after one week. This technique provides additional working convenience during the laying of slabs. Provision of construction joints is easier.

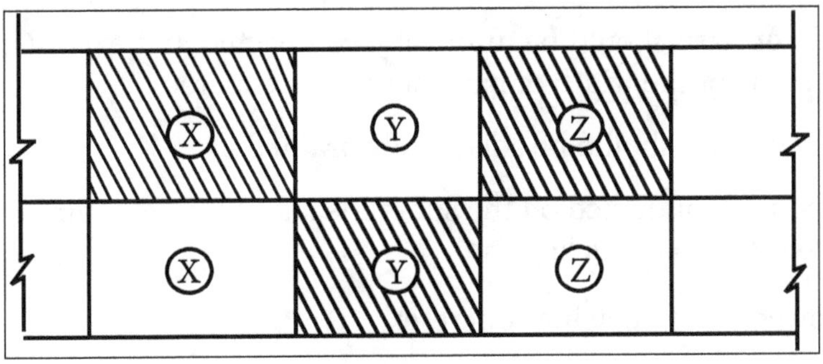

Modes of construction of cement concrete road.

This mode of construction has the following setbacks:

- More number of transverse joints have to be provided and thereby increasing the cost.
- Possibility of collection of surface water on the base or subgrade and thereby disturbing the base or subgrade.
- Diversion of traffic is needed as the construction is done on alternate bays covering the entire width.

2. Continuous Bay Method

In the continuous bay method, X, Y, Z, etc. are done at a stretch in sequence. Construction joints are however provided at the end of day's job. In general the second method is preferred as construction of one half of the pavement can be constructed while the other half is being used by traffic.

Construction Procedure of Pavement Slab

Sequence of construction of the pavement slab consists of the following steps:

- Preparation of subgrade and base.
- Placing of forms.
- Installation of joints.
- Batching of aggregates and cement.
- Mixing and placing concrete.
- Consolidation and finishing concrete.
- Cutting of concrete.

1. Preparation of Subgrade and Base

The subgrade or base should be prepared complying with the following conditions:

- No soft spots are present in the subgrade or base.
- Subgrade or base should be uniformly compacted and extended about 30 cm, on either side of the width of pavement to be concreted.
- Subgrade or base should be adequately drained.
- Plate load test conducted on the subgrade should yield a minimum modulus of subgrade reaction of 5.5 kg/cm3.

Steps to be followed before placing concrete:

- The subgrade is prepared and checked atleast two days in advance of concreting.

- The subgrade or base is kept in moist condition at the time of concrete so as to avoid the subgrade absorbing water from concrete. If necessary, the subgrade may be saturated well with water for 6 to 20 hours prior to placing of concrete.
- Water proof paper may be spread on subgrade if concrete to be placed directly on the subgrade. In this case no saturation of subgrade is necessary.

2. Placing of Forms

Wooden or steel forms are used. Wooden forms have minimum base width of 10 cm for 20 cm slab thickness and of 15 cm for slabs over 20 cm thicknesses. Forms are joined neatly and are set with exact grade and alignment. Forms are rigidly fixed such that during the entire operation of concreting they should not deviate more than 3 mm from straight edge of 3 m length.

Steel forms commonly used are straight 3 m sections. They are aligned vertically and horizontally by slip joints and held in position by three or more steel stakes. These stakes are of 2.5 cm diameter, 45 to 75 cm long driven at intervals at back of form. Forms are available in heights of 15 to 30 cm and with corresponding base width over a similar range. These are made of 4 to 6 mm steel plate.

3. Installation of Joints

Extreme care should be taken in all operations connected with joints. Face of transverse joints should be straight, perpendicular to the center line of pavement and also perpendicular to the surface of the finished slab. Load transfer devices like dowel bars used in expansion joint should be aligned and placed accurately. There should be free movement of slab ends in longitudinal direction.

4. Batching of Aggregates and Cement

Based on the design concrete mix, the proportion of ingredients like coarse aggregate and fine aggregate are proportioned by weight in a weight batching plant. These are placed in the hopper of the mixer along with the necessary quantity of cement.

Cement is measured by the bag which measures 50 kg. All batching of material is done on the basis of one or more whole bags of cement taking the unit weight of cement as 1440 kg/m3. The mixing is done in batch mixer.

Mixer performs well such that there is uniform distribution of materials and mixed material is uniform in color and homogeneous. The batch of cement, fine aggregate and coarse aggregates are added to the mixer. The water for mixing is introduced into the drum within the first 15 seconds of mixing. After all the materials are placed in the mixer, the mixing is to be started within one- and-a half minutes.

5. Mixing and Placing Concrete

The ingredients are mixed in required quantity for immediate use and are deposited on the subgrade or base. The concrete should be deposited to the correct depth and width of pavement section within the formwork. The operation of placing concrete should be continuous. Care should be taken to see that no segregation of materials takes place in any stage of the operation.

6. Consolidation and Finishing

Concrete is spread uniformly by shovels with re-distribution wherever needed. Needle vibrator is used for compaction. Surface of the pavement is compacted either using a power-driven finishing machine or using a vibrating hard screen. Hard compaction may be resorted to where the width of slab is very small and at corners.

After placing and operating concrete, the crown and the cross section of the pavement has to be done with the required grade. Concrete is further compacted parallel to carriageway center line by using a longitudinal float.

After the longitudinal floating is done and the excess water gets disappeared the final grade and level are tested with the straight edge. Before the concrete gets hardened, the surface is belted with a two-ply canvas belt.

After belting, the pavement is given a broom finish with fiber broom brush. Before the concrete develops initial set, the edges of the slab are finished with edging tool.

7. Curing of Concrete

It is very important to ensure proper curing of the finished concrete.

Following are the methods usually adopted:

- Earth cover kept wet.
- Hay or straw cover kept wet.
- Cover of wet felt mats or cotton mats.
- Saw dust kept wet.
- Continuous sprinkling of water.
- Surface application of calcium chloride or sodium sulphate.
- Cover by water-proof paper or impervious membrane.

Forms are not removed until concrete has set for atleast 12 hours. Special devices are used to pull the forms without damaging the pavement. At the end of curing period the surface grooves in longitudinal and transverse joints are cleaned and filled with sealing compound. After the clearing the road surface, the road is open to traffic.

4.5 Polymer Modified Bitumen

Styrene-butadiene-styrene block copolymer (SBS) is the most widely accepted bitumen/asphalt binder modification agent used by the whole bitumen industry.

Improvement in rutting resistance, thermal cracking, fatigue damage, stripping and temperature susceptibility have led SBS polymer modified binders to be a substitute for non-modified, neat bitumen.

Applications in paving and maintenance include the following:

- Hot mix,
- Warm mix,
- Cold mix,
- Chip seals,
- Hot and cold crack filling,
- Patching,
- Recycling,
- Slurry seal.

They are used wherever extra performance and durability are desired. In many cases, they are selected to reduce life cycle costs. Polymer modified binders have allowed the use of paving techniques previously not viable such as micro-surfacing and use of emulsion chip seals on high traffic roads.

The rheological behavior of SBS polymer-bitumen blends is of great interest because it is closely related to the performance of pavements. Addition of small amounts of polymer dramatically changes the properties of the asphalt binder. Addition of polymer introduces an additional difficulty to the bitumen complexity, because of its high molecular weight.

Formation of a homogeneous system is not easy and incompatibility problems arise. After the manufacturing of the polymer-bitumen blends, phase separation may eventually occur as the blend is stored at high temperature (160–200 degree Celsius) in absence of stirring. Pavement construction techniques require the binder to remain stable previously and during the application on the road.

Prior to utilization of SBS polymer modified bitumen into various countries it requires careful planning, engineering design and many laboratory tests in order to;

- Maximize benefits,
- Minimize cost,
- Perfectly accommodate into the region's individuality.

It is easy to understand that:

- The SBS polymer modified binder and pavement should fulfill the local traffic and climate requirements.
- Raw material properties and prices (mostly bitumen/asphalt binder) have significant differences in different regions/countries because of various refinery processes.
- Bitumen – polymer compatibility has to be understood in order to minimize phase separation.
- Usually the performance-cost optimum PMB has to be created not the highest performance one.

Objective of polymer-bitumen technology transfer:

The main objective of SBS polymer modified bitumen technology transfer is:

To suit the binder and pavement design, Construction experience to local needs in order to have a successful adaptation.

Common steps of a complete technology transfer:

Before introducing SBS copolymer modified bitumen/asphalt into a new market in general the following steps suggested to be followed and accomplished.

General Study

The main goal of a general study is to inform decision makers and anybody who needs introduction level information on PMBs.

Technical Study

The technical study discusses the following:

- Raw materials.
- Compatibility issues.
- Manufacturing methods.
- Mixing conditions.

Feasibility Study

It presents detailed economic indicators, taking the capital cost and detailed operating costs into considerations. Comparison with production cost of today's market leads to modified bitumens.

Binder Design and Formulation

Synchronized formula design and laboratory work lead to the development of PmB formulas having different performance and price. The time-frame of the development depends on desired quantity of PmB formulas.

Specification Development for PmB Formulas

Adaptation of existing standards/specifications can be an option, but without customizing to the local needs it might be too general. It has many values if local standard is developed but not adopted.

If local standards are not available bitumen engineering can contribute in specification development and/or define the key process parameters which are necessary for the required performance and result in minimal cost in quality control.

Standardization of Formulated Binders

This step contains preparation of complete technical documentation for local agencies, specimen production and evaluation of SBS polymer modified bitumen binders.

Mix Design and Asphalt Mixture Tests

Modified binder formula is not viable unless it is tested in asphalt mixtures.

This section covers the complete laboratory testing and Field evaluation of the PmB formulas. Various asphalt mixture tests are performed to evaluate the performance according to existing local, national standards and specifications.

Design and Manufacturing or Purchase of the Blending Unit

Bitumen Engineering recommends considering manufacturing of other modified bitumens as alternative options. This provides maximum flexibility in responding to raw material price changes, market needs etc.

There are three main options here:

- Bitumen Engineering can design a blending unit explaining the clients all recommended components. Then recommend manufacturers and work with the chosen one until start up.
- Bitumen Engineering recommends equipment suppliers.

- Bitumen Engineering does not contribute in plant design and manufacturing but provides all information obtained from the client who deals with choosing thc appropriate vendors/equipment manufacturers.

4.6 Recycling

It is estimated that roadway and highway construction in the United States is currently consuming between 500 and 800 million metric tons of virgin crushed rock, gravel and sand each year as aggregate.

The existing annual quantities of industrial by-product source materials represents a way to reduce the amount of virgin materials required for transportation of infrastructure projects, natural resources, address energy and climate change issues and reduce the need for landfill space.

The industrial by-products are divided into six major groups as follows:

- Coal Combustion Products.
- Iron and Steel Slag Products.
- Foundry Slags and Sands.
- Tire Derived Rubber Materials.
- Recycled Concrete Materials.
- Paper Production Materials.

Recycling.

Coal Combustion Products

Coal combustion products (CCPs) are the byproducts generated from burning coal in coal-fired power plants. These byproducts include coal fly ash, bottom ash, boiler slag and flue gas desulfurization gypsum.

Benefits of using CCPs

Proper use of CCPs in building applications can yield environmental, economic and product performance benefits. Using CCPs in an environmentally safe manner will save virgin resources and reduce energy consumption and greenhouse gas emissions (GHG). It will help reduce the need for landfill space and new landfills. The use of CCPs also makes good economic sense since they are often less costly than the materials they replace.

Environmental Benefits and Greenhouse Gas Benefits

There are many ways in which the reuse of CCPs reduces the generation of GHG.

For example, using existing CCP materials in place of virgin materials reduces energy intensive mining operations needed to generate virgin materials and the resulting emissions and fuel consumption that result in GHG.

The primary GHG reductions associated with using coal ash come from its replacement of Portland cement in concrete. Portland cement is very energy intensive to manufacture. It takes 55 gallons of oil to produce a single ton of cement.

The pozzolanic properties of coal fly ash make it a useful replacement for a portion of the Portland cement used in making concrete. Fly ash can typically replace between 15 to 30 percent of the cement in concrete with even higher percentages used for mass concrete placements. In addition, it makes the concrete stronger and more durable than concrete made with only Portland cement as the binder.

Benefits from Reducing the Landfilling of CCPs

Using CCPs instead of landfilling them reduces the need for additional landfill space. The landfill space required is the equivalent of placing 26,240 quarter acre home-sites under 8 ft. of CCPs. Many energy facilities no longer have adequate storage space for CCPs.

Using CCPs reduces the need to locate and develop new disposal facilities and any adverse environmental or health effects associated with them.

Benefits from Reducing the Need to Mine Virgin Materials

CCPs can be substituted for many virgin materials that would otherwise have to be mined. It includes, lime to make concrete, natural gypsum for making wallboard and gravel for making roofing granules. Each of these materials would require mining virgin materials, potentially destroying green fields and wildlife habitat.

It makes more sense to use existing materials that would otherwise be disposed of than to mine new ones. At the same time, waste and harm to the environment is reduced.

Performance and Economic Benefits

Each type of CCP has particular performance benefits. For example, coal ash can be used to create superior products because of its inherent properties. Mixing fly ash with Portland cement mixtures can produce stronger and longer lasting buildings than concrete made with only Portland cement as the binder (glue).

This not only reduces costs of maintaining buildings but provides the additional environmental benefit of reducing the need for new concrete to replace aging buildings. This leads to a significant reduction in future energy consumption and GHG emissions.

Boiler slag, which replaces sand in blasting grit, has the benefit of being free of silica which eliminates the potential health risk of silicosis.

4.7 Different Materials

Glass and its Constituents

Glass is a transparent or translucent material which is hard and brittle. It is a substance or combination of substances which have solidified from the liquid state without crystallization. Main constituents of glass are silica, sodium or potassium carbonate, lime, manganese dioxide, cullet and colouring materials.

1. Silica

It is the principal constituent of glass. When fused at a very high temperature would give a quality glass. To make the molten silica glass sufficiently viscous, it is essential to add some quantity of alkaline material. Such addition is done so as to get good workability and resistance against weathering agencies.

2. Sodium or Potassium Carbonate

Sodium or potassium carbonate is an alkaline material. It is added with a certain proportion so as to lower the melting point of silica and to impart viscosity to the molten glass.

3. Lime

Lime is added in the form of chalk so as to impart durability to glass. To improve the brightness and shining lead oxide is added in the place of lime.

4. Manganese Oxide

To correct the colour of the glass due to the presence of iron in raw materials of glass, manganese oxide is added in certain proportion.

5. Cullet

Cullet is the old broken glass of the same type as that one is intended to be manufactured.

6. Colouring Materials

In the process of manufacturing coloured glass, necessary suitable colouring substance is added at the fusion stage in order to provide the derived colour of the glass.

7. Properties of Glass

Properties of glass are based on the constituents, surface condition, thermal treatment, dimensions of the object, etc.

Following are some of the important properties:

- It has no defined crystalline structure and has no sharp melting point.
- It is capable to absorb, refract or transmit light.
- At elevated temperatures it acts as an excellent electrical insulator.
- It is not affected by air or water.
- It cannot be attacked by ordinary chemical agents.
- It can be altered to meet certain requirements by changing fusibility, hardness, refractive power, etc.
- It is possible to weld pieces of glass by fusion.
- Based on the advanced technology it is possible to make glass lighter than cork or softer than cork.
- It is affected by alkalies.
- The only drawback is that it is brittle.

Fibers

1. Natural Fiber and its Properties

Naturally available fibrous material is asbestos. Main constituents of asbestos are silicate of calcium and magnesium. It contains a small amount of iron oxide and alumina. Original source of asbestos was the mineral fibrous mineral actinolite.

Belongs to two groups, via, those which are readily active with acid (i.e., nonacid-resistant) and others are of acid-resistant. The first group consists primarily of chrysolite asbestos which is used for industrial purposes. The second group comprises of amorsite asbestos and crocidolite.

Second group is not recommended for commercial use as they are believed to cause diseases related to lungs. Indian industry uses the first group asbestos.

Following are the properties of asbestos:

- It has high tensile strength along the flame.
- It is incombustible and hence acts as a good insulator for heat.
- It is acid proof.
- It is flexible, soft and non-porous.
- It is fire-resistant and the welding point is 1200°C to 1500°C.
- It possesses good adhesion with cement and hence the fibre can be used as a fibre reinforcement.
- It is rust free.
- Its quality is critically affected by the length of fibres.
- It can be easily cut into pieces.

2. Reinforcing Fibres

Composite materials are formed by combining two or more different materials to utilise their virtues and minimizing their deficiencies. These materials are called composites. Depending on the requirement, higher and lighter materials may be combined.

In the finished product, the materials used retain its physical and chemical properties independently. Compacts comprise of two main constituent materials.

- Strong load carrying material called as reinforcement or reinforcing fibres.
- Weaker materials are known as matrix.

The functions of reinforcing fibres are:

- It provides rigidity and strength.
- It supports the structural loads.

Three most commonly used reinforcing fibres are,

- Glass fibres,
- Carbon,
- Aramid.

Glass fibres are the heaviest which provide high flexibility.

Carbon is the strongest and heavier.

Aramid has moderate stiffness.

When more than one reinforcing fibre is used, that material is called hybrid composites.

Plastics

Organic substances which may be natural or synthetic resins in origin with or without moulding components. Synthetic resins are: phenol, formaldehyde, cellulose, vinyl, alkyd, etc. The moulding compounds are catalysts, filters, hardners, lubricants, pigments, plasticizers, solvents, etc. Based on their behaviour with respect to heat, they may be broadly grouped as Thermoplastic and Thermo-set plastics.

1. Thermoplastics and their Properties

Thermoplastics become soft when heated and become hard when cooled. For indefinite time period, these plastics can sustain repeated softening and hardening. Old scrap and worn-out materials of this type can be effectively used again.

Some of the thermoplastics which are of important are: Acrylic, Cellulose Acetate, Polythene, Perspex and Polyvinyl chlorides. Acrylics are transparent, rough and strong which can withstand impacts. Cellulose acetate is made from cotton seed.

Polythene is transparent and chemically unaffected by temperature and moisture. Perspex is strong enough and is made in the form of sheets. Poly Vinyl Chloride is a product obtained from vinyl chloride and acetate. It is acid proof, light and withstands wear and tear.

2. Thermoset Plastics and their Properties

It becomes rigid due to chemical changes, when moulded under pressure and temperature. These plastics once cooled and taken a form cannot be changed further due to the application of heat. Some of the commonly used thermoset plastics are phenol formaldehyde, phenol forfuraldehyde, urea formaldehyde and bakelite.

Phenol formaldehyde is formed by heating phenol and benzene with a suitable catalyst. Urea formaldehyde is formed by reacting urea with formaldehyde in the presence of a catalyst. Bakelite is made by heating equal amount of phenol and formaldehyde with a base.

Geo-textile Subgrade System

A geo-textile with good tensile strength can contribute to the load carrying- ability of the soil, which is poor in tension and good in compression. The prime function of geo-textiles in unpaved road or weak soils is to separate aggregate from the subgrade and then act as a reinforcing medium. The two principal mechanisms that appear to contribute to the reinforcement function of a geo-textile are:

- The ability of the fabric to confine and restrain moment of the granular structural layer.
- The membrane effect whereby a fabric that develops high tensile strength under load can induce a significant vertical stress upward.

This helps the granular layer to support the vehicular load and reduce stress intensity on the subgrade.

Geo-membranes

Geo-membranes are thin materials with very low permeability. They are flexible and are manufactured from synthetic or bituminous products. They may be strengthened if necessary, with a fabric or film. Geo-membranes differ from geo-textiles with reference to the rate of permeability. The permeability is high in geo-textiles and very low in geo-membranes.

For all practical purposes, geo-membranes may be considered to be impermeable to both gases and fluids. It is ideal for constructing waterproof or gas proof barriers between adjacent bodies of soil or soil and fluid. Geo-membranes are used in the following situations:

- Sealing against fluid percolation.
- Buffers against pollutants.

Geo-membranes are manufactured from synthetic (thermoplastic) or bituminous products. Some of the synthetic materials used are high-density polyethylene, low-density polythylene, polvinyl chloride.

4.8 Quality Control Measures

Quality control of subgrade, quality of materials used for base, sub-base and weathering and concrete are to be made.

Quality Control of Subgrade

It is considered as a part of the pavement to a depth of about lm. Subgrade should have strength and stability depending on the type of pavement and traffic. Quality of the subgrade is checked by conducting a field CBR test and plate bearing test. If the subgrade is of inadequate strength, quality and stability required, it may have to be improved.

The improvement may be brought by improved compaction, lime or cement stabilization, excavating poor material and replacing with select materials, all adequate compacted and or by extra drainage. If the subgrade is compacted it should have 90% of max dry density as obtained from Proctor's compaction test. Necessary in-place density test, core-cutter or sand replacement method may be adopted.

Quality Control of Bases and Sub-bases

Base courses are also used under rigid pavement for preventing pumping and protecting the subgrade against frost action. Suitable strength and stability tests like plate bearing, CBR or stabilometer tests are conducted. Quality requirements for each material used should be ensured.

Quality Control of Wearing Courses

The prime aim of providing wearing course is to give a smooth riding surface which should evidently be dense. In general for flexible pavements a bituminous surfacing and for rigid pavements cement concrete surfacing are used. These wearing courses apart from strength and stability requirements should act as a water tight layer. As such there is no test available for evaluating the structural stability. But the bituminous mixes used in the wearing courses are tested for their suitability.

Marshall Stability test is used to find the stability and density. Plate bearing and Bankelman beam tests are used to evaluate the overall strength and deflection. All the parameters should be as per standard to ensure quality.

Quality Control of Flexible Pavements

A typical flexible pavement consists of four components:

- Soil subgrade,
- Sub-base course,
- Base course,
- Surface course.

A well compacted granular structure consisting of strong graded aggregate can transfer the compressive stresses through a wide area and thus form a good flexible pavement layer. The load transfer capability evidently depends on the type of the materials and the mix design factors.

Bituminous concrete is one of the best flexible pavement layer materials. Other materials which fall under this group include all granular materials with or without bituminous binder; Water bound Macadam, crushed aggregate, etc.

Bituminous concrete or Asphaltic concrete is a dense graded premixed bituminous mix which is well compacted to form a high quality flexible pavement. It consists of proportioned mixture of coarse aggregates, fine aggregates, mineral filler and bitumen.

The mix is designed based on Marshall Method to fulfill the requirements of stability, density, flexibility and voids. Necessary quality tests are made at every stage of construction based on the material used for flexible pavement.

Quality Control of Rigid Pavements

The concretes placed for rigid pavements have to be cured for a period of 7 to 25 days in order to allow the concrete to set. Curing may be done as per the standard procedures. Following quality control of concrete has to be done:

- Two specimens for every 150 cubic cm have to be taken for cube strength and compressive strengths at 7 days and 28 days of curing have to be determined.
- Cores should be taken to check the density during the curing period periodically.
- Horizontal alignment should be checked frequently.
- Quality of cement, aggregates and water are to be checked periodically.
- A slump test should be conducted frequently during the process of lying.

Quality Control of Rural Roads

In general, quality control in construction of roads in rural areas is not of the same standards as for higher types of roads. It becomes necessary to have modified quality control measures so as to meet the minimum requirements. Earth and water bound Macadam are in general constructed with sub-base, base and wearing courses.

1. Earthwork

Generally the volume and composition of traffic are less in rural roads and acceptable level of serviceability is less. Keeping these factors it is considered sufficient to have 30 cm of the road formation as subgrade. Moisture and dry density requirements followed in higher types of roads need not be strictly followed in rural road construction.

Instead, it is sufficient to ensure that the moisture content is kept within ± 2.5% of OMC for clay and clayey soils and ± 1.5% of OMC for other soils. As regards to in-situ densities it may be 0.1 gm/cc less than the specified densities for fine grained soils and 0.15 gm/cc for coarse grained soils. Tractor towed/animal drawn road rollers may be made use of in place of power driven road rollers. However, the lift thickness will be less and the number of passes will be more.

2. Sub-base and Base courses

The gradation and plasticity requirements for sub-bases and base causes are given below:

(1) Gradation Requirements

Fraction Percentage (by weight)

- Passing 40 mm size and retained on 2.35 mm sieve 45-65.
- Passing 2.36 mm size and retained on 75 micron sieve 30-40.
- Passing 75 micron sieve 5-15.

(2) Plasticity Index

Plasticity index of fine fraction (Passing 425 micron) should be less than 6.

3. Wearing Course

The gradation and plasticity index for wearing courses are given below:

(1) Gradation Requirements

S. No.	Fraction	Percentage (by Weight)
(i)	Passing 40 mm size and retained on 2.36 m sieve	30-60
(ii)	Passing 2.36 mm size and retained on 75 micron sieve	30-50
(iii)	Passing 75 micron sieve	10-25

(2) Plasticity Index

Plasticity requirements modified to suit the climatic conditions for the wearing course are as given below:

Climatic Condition		Liquid Limit Not to Exceed	Plasticity Index Range
(i)	Moist temperate and wet tropical	35	4-9
(ii)	Seasonal wet tropical	40	6-15
(iii)	Arid	55	15-30

Mechanical stabilization is more appropriate for compacting sub-bases and bases for rural roads. If cement/lime stabilization is resorted to uniformity of mixing then the coefficient of variation in the distribution of stabilizer content should be ≤ 15%. If the aggregate Impact Value testing equipment is not available, the quality of aggregates is assessed through the moisture absorption of the aggregates.

The recommended limits of water absorbed for different pavement layers are given below:

Layer	Water Absorption (Percent) Not more than
Sub-base course	6
Base course	3
Wearing course	1.5

2. Water Bound Macadam

Recommended gradings for stone aggregates in WBM layers may be simplified for rural roads as given below:

Percentage Passing (by Weight)

Sieve Size	Grading I(90-40mm)	Grading II(63-40mm)	Grading III(50-20mm)
100 mm	90-100	—	—
63 mm	35-70	90-100	—
50 mm	—	35-70	90-100
40 mm	0-10	0-10	35-70
20 mm	—	—	0-10

The plasticity index requirements for the fill material of WBM should be the same as for sub-base/base/wearing courses. The amount of water to be adopted depends on the type of soil on which the road is constructed and the type of stabilizer used.

4.9 Highway Drainage

One of the major causes of failure of highways is presence of water. A part of rain water falling on road surface and the surrounding area is lost by evaporation and percolation.

Remaining water known as surface water either stays on the surface of the road and the surrounding area or flows away depending on the topography of the area. Removal and diversion of surface water from the road and surrounding area is termed as surface drainage. Diversion or removal of excess water beneath the ground surface is termed as sub-surface drainage.

Drainage Effects on Highway

Improper drainage facility of a road causes destruction in the following manner:

- Road surface becomes soft and losses strength if the pavement is of soil, gravels or water bound macadam.
- Bearing capacity of subgrade soil is reduced because of presence of water.
- If the subgrade is of expansive soil, variation in moisture causes the subgrade soil to expand or contract.

- Formation slopes may fail due to poor drainage.
- Rainwater has to drain along the road side failing which slips and landslides may occur causing road failures.
- Improper maintenance of drainage may cause erosion of side slopes, side drains and formation of gullies.
- In sub-zero weather conditions the water in the subgrade gets frozen and because of this the ice in the voids expands resulting in heave of pavement.

Requirements of Highway Drainage

For the best performance of the road the following requirements have to be fulfilled:

- Surface water should not be permitted to stagnate on the road surface and shoulders. Arrangements should be made to drain off this water.
- The water collected at the adjoining area due to rain should not be allowed to flow towards the road. Adequate slope should be made for the adjoining area.
- Side drains should be provided with sufficient longitudinal slope so that it may drain off the collected surface water effectively.
- Surface water flowing across the pavement should not develop cross ruts or erosions on road surface and shoulders. For this high embankment slopes should be protected either by turfing or pitching.
- Ground water level should be maintained well below the road surface. The water table should be 1 to 2 m below the road subgrade.
- Special steps should be taken to keep down the harmful salts in water logged areas.
- Road formation should be well above the high flood level, atleast 10 cm above high flood level.

Surface Drainage

It deals with arrangements for quickly and effectively leading away the water that collects on the surface of pavements, shoulders and other adjoining areas.

Surface drainage consists of the following two operations:

- Collection of surface water.
- Disposal of collected surface water.

Rain water from road surface is let off to the sides by cross-slope or camber. Based on the rainfall of the area the rate of cross-slope is provided. In rural plain area, the disposal of water depends on whether the road is in embankment or in cutting or on ground line.

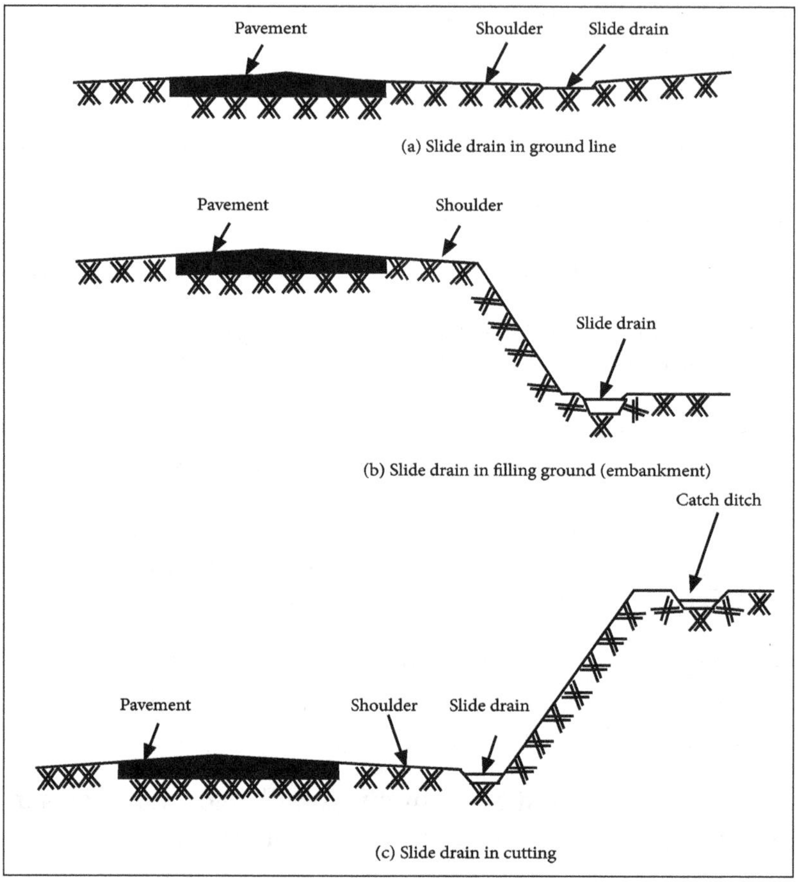

Arrangement of Side-Drains.

Side drains of suitable size and of adequate longitudinal slope are constructed along both the sides of the road at some distance from the foot of embankment. The side-drains are trapezoidal in shape. In cuttings, drains are provided along both the sides of the road just next to the shoulders. If there is a restriction of space covered drains or drainage trenches are provided with layers of coarse sand and gravel, as shown in Figure.

In case of urban roads, underground drainage facilities are provided to drain off surface water. Urban drains are provided, as shown in Figure, because of presence of foot path and other developments. Water drained from the pavement surface can be drained longitudinally and may be collected in catch pits and carried forward through underground drainage pipes. A typical catch pit with grating to prevent the entry of rubbish into the drainage system is shown in Figure.

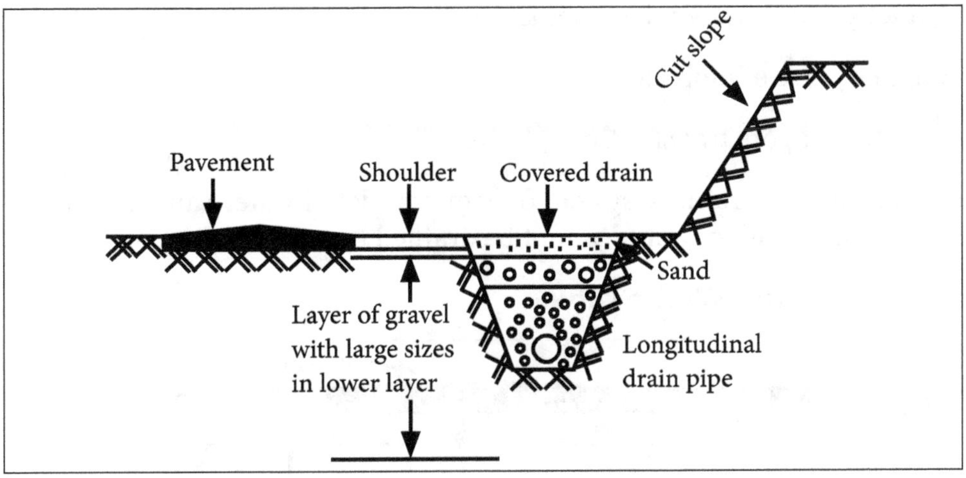

Trench filled with gravel and sand.

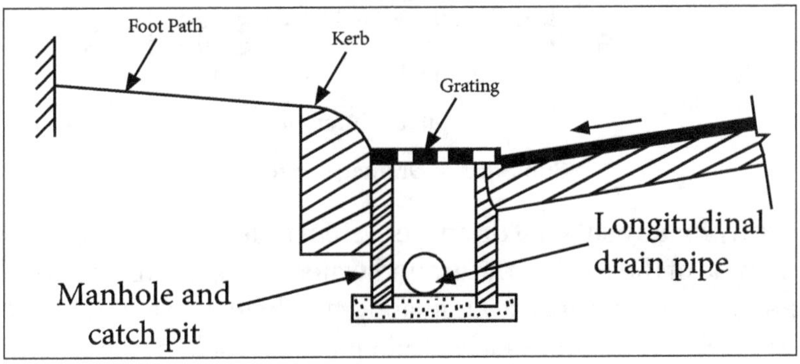

Surface Drainage system in urban roads.

Highway drainage is of much more important in hill roads. In addition to the drainage of water from the road formation, it is essential to divert and dispose of the water flowing down the hill slope across the road and that from numerous cross streams. If the drainage system in hill road is not adequate and efficient it will result in complex maintenance problems. Hence drainage arrangements in hill road should be made to work efficiently.

Sub-Surface Drainage

Stability and strength of subgrade is the important aspect to be considered for the best performance of roads. Strength of subgrade is decreased due to the presence of water. It is better if the subgrade is always kept dry.

Variations in moisture content of subgrade are caused by the following:

- Seepage of water from higher adjoining ground.
- Penetration of moisture through pavement.
- Percolation of water from shoulders, pavement edges and soil formation slopes.

- Under ground water table fluctuation.
- Capillary size of moisture.
- Moisture vapors transfer through the soil.

By providing sub-soil drainage system only gravitational water can be controlled. But water vapour and capillary water cannot be drained by sub-soil drainage system.

1. Control of Ground Water Table

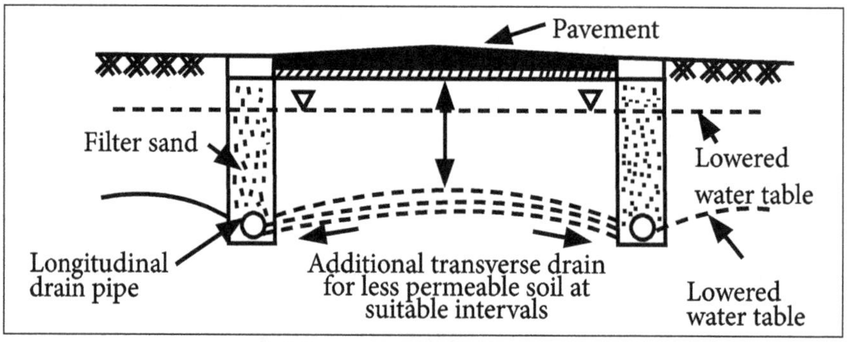

Sub-Surface Drainage system.

There is no necessity for any sub-soil drainage system if the ground water is deeper than 1.5 m below the subgrade of the road. If the ground water is closer than this, the best approach is such that the subgrade rests atleast 1.2 m above the highest water table. If the above approach is not feasible and subgrade soil is drainable, deep side trenches are constructed to lower the water table. Figure shows the arrangements of drain pipe on filter sand.

If the soil is relatively less permeable, transverse drains may have to be provided (shown dotted in Figure.)

2. Control of Seepage Flow

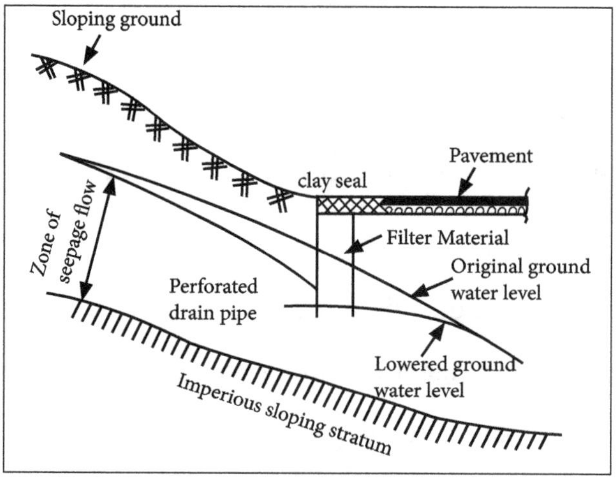

Control of Seepage Flow.

Seepage flow is likely to exist along a sloping ground. If the seeping zone is at a depth less than 0.6 to 0.9 meter from the road surface, it is necessary to intercept the seepage flow. The arrangement of sub-surface drain in such a case is shown in below Figure.

3. Control of Capillary Rise

The subgrade is likely to be disturbed by capillary water if the subgrade is near the capillary zone. In this case instead of lowering the water table, the capillary rise can be arrested. The capillary rise can be checked by one of the two methods:

- By providing a granular material of suitable thickness during construction of embankment between the subgrade and the highest level of sub-surface water.
- By providing a capillary cut-off by inserting an impermeable or a bituminous layer in the place of granular blanket.

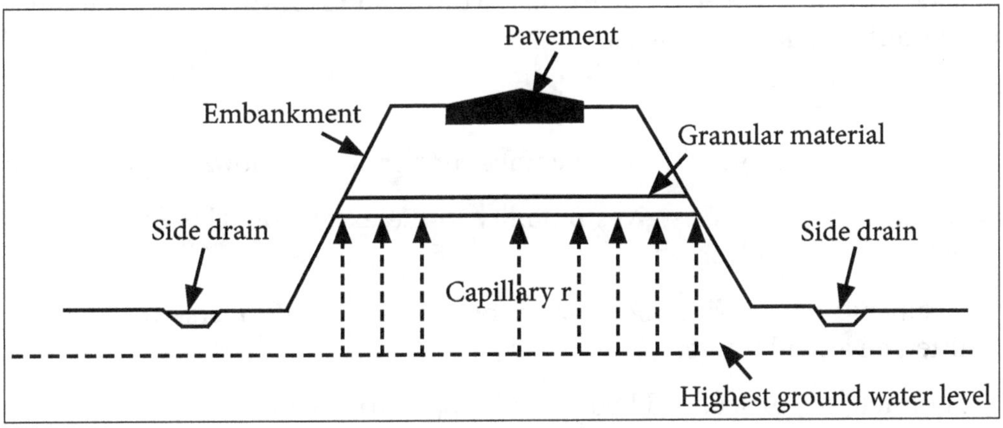

Granular Capillary Cut-off.

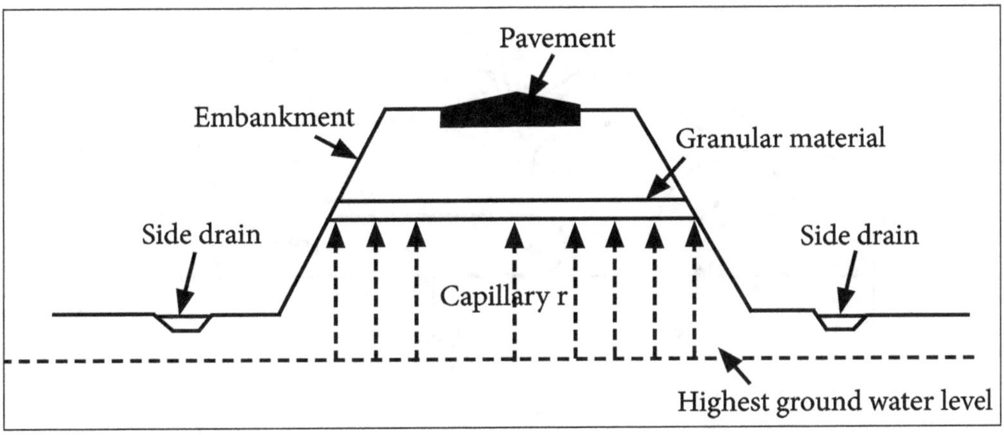

Impermeable Capillary Cut-off.

4.10 Construction Machineries

Construction machineries are needed for a highway construction project for the following:

- Earthwork.
- Preparation of sub-grade.
- Embankment construction (if needed).
- Construction of pavement structure.

Earthwork Equipment and Operation

Wide range of excavating equipment is available. It demands a much greater care and thought in selecting the most suitable machine for a particular job. No machine will exactly satisfy the 100% need of the work. Earthwork and moving equipment are available to suit different field condition.

1. Excavators

These are digging machines. These machines consist of the following components:

- An undercarriage-This gives mobility to the excavator. This may be mounted with crawler track or wheel.
- A superstructure with operator's cabin - This could traverse through 360° or fitted on a rigid frame.
- Hydraulically articulated booms and dipper arms with bucket.

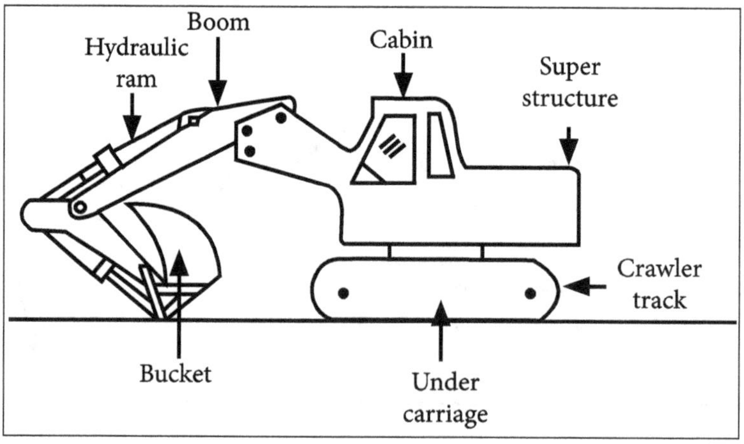

Excavator

Excavators include dipper or power shovel, dragline, clamshell and drag shovel. All these fall under shovel family in that the front end only is changed. Shovel based machines have a common type of operating and tracking mechanism. The excavators may be of mechanical or cable operated or hydraulic types. Hydraulic excavators have more advantages.

They can be fitted with various multipurpose attachments for various earthmoving works. Most efficient action of the machine is that the hydraulic-unit of action of the excavator directs the bucket teeth at their most efficient angle during dipping operation. Mechanical or cable operated excavators are of either hoe or shovel type. In hoe type, the bucket opening faces towards the machine. In shovel, the bucket opening faces away from the machine.

2. Shovels

It is used for a specific type of excavating machine fitted with a short-length boom and working with forward strokes.

Following are the four types of machines which have the same basic structure and the front end attachments are different:

- Dipper-shovel,
- Drag shovel or hoe,
- Dragline,
- Clamsheel.

The front line attachments are:

- Shovel boom and dipper stick for a dipper shovel.
- Shovel boom, jack boom and stick for a drag shovel.
- Crane boom with a fairlead and a loosely attached bucket.
- Crane boom with a special type of bucket for a clam sheel.

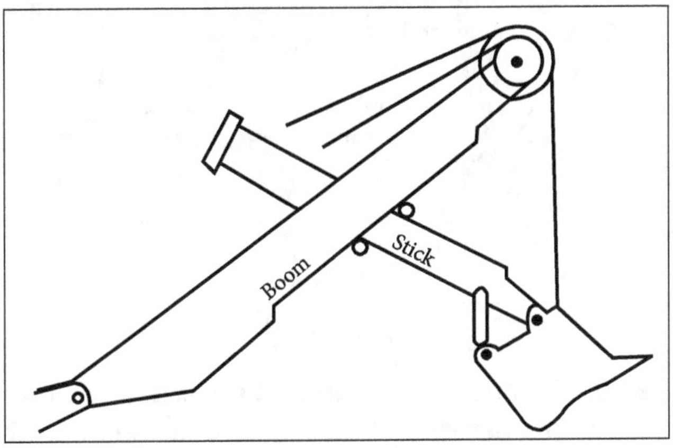

Attachment for dipper shovel.

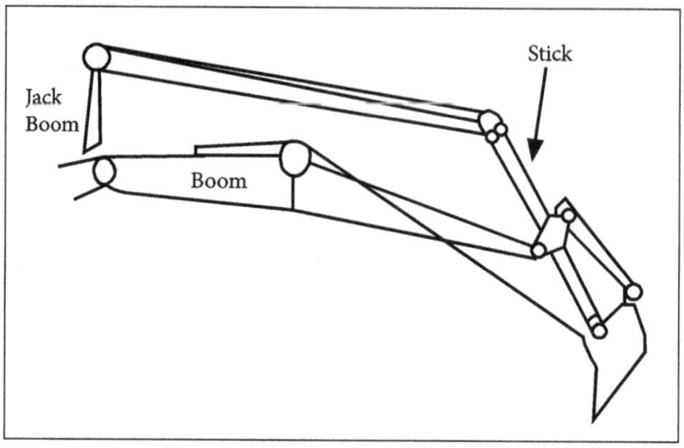

Attachment for a drag shovel or hoe.

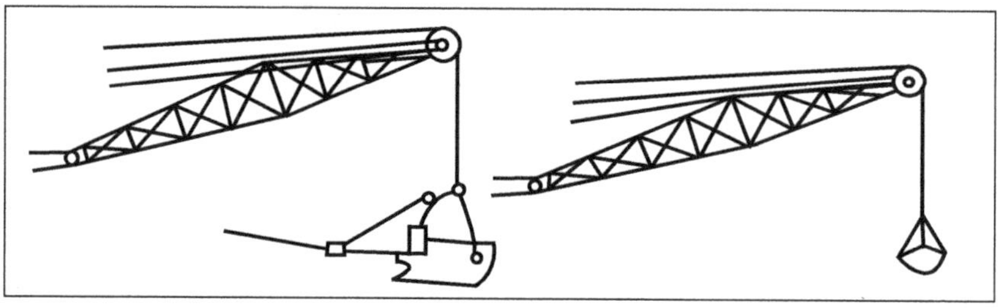

Attachment for a dragline Attachment for a clamshell.

(i) Dipper Shovel

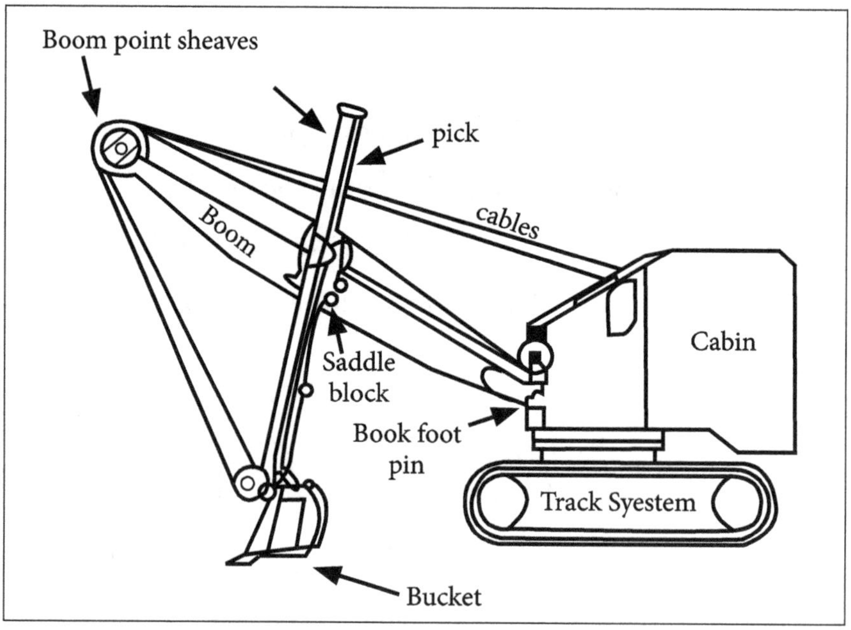

Dipper shovel.

It is the most popular of the shovel excavator. Consists of a boom, dipper stick, the bucket and the mechanism of operation. Lower end of the boom is hinged to the support bracket and the upper end carries the bucket hoist sheaves. Dipper stick moves back and forth on a dipper drum through a guide formed by a saddle block.

In order to cause the stick to slide back and forth the dipper drum is rotated. The bucket of the power shovel is strongly built of steel. Generally used for digging hard rock in quarries or ores in mines. Digging is done above the machine base level.

(ii) Drag Shovel or Hoe

It is also called as a Pull shovel. This machine resembles the digging mechanism of an ordinary garden hoe from which it gets the name. The digging action results from the drag or pull of the bucket towards the machine. This action helps for digging below the machine level particularly while digging for trenches, footings, basements, etc.

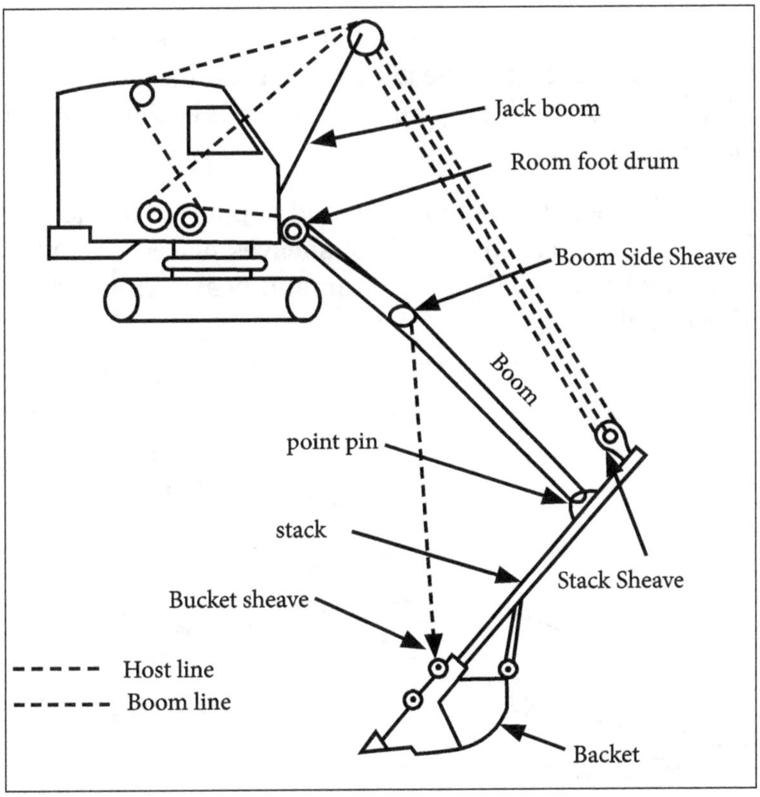

Drag shovel.

Shovel consists of a boom, a stick, a jack boom and a bucket which is attached at the end of the stick. Stick in hoe is hinged by a pin with the boom. It enables to turn over the pin to take any desired direction suited for digging or dumping operations. The lower end of the stick carries the bucket and the upper end carries a sheave having the hoist cable supported on a jack bottom at the other end. This arrangement enables to hoist the cable when pulled and the reaction at the stick enables to move the boom with stick up or down. Thus the boom is capable of taking any position in vertical plane.

Applications

Drag shovels have application in the following:

- Digging trenches, footings or basements.
- Digging for hard materials.
- Providing close trimming whenever needed.
- Used where excavated material needed to be dumped at a short distance.

(iii) Dragline

As this machine drags the bucket against the material to be dug it attains this name. It consists of a boom, a bucket, a fairlead and hoist, dump and drag cables. Boom is similar to that of a crane boom. The top and bottom ends of the boom are known as the point and the foot respectively.

Boom is attached to the deck with pin at foot and supported at the point through the boom hoist cable. Cable passes over a pair of sheaves placed one on each side. They have three types of buckets, viz., light, medium and heavy which are used depending on the nature of material.

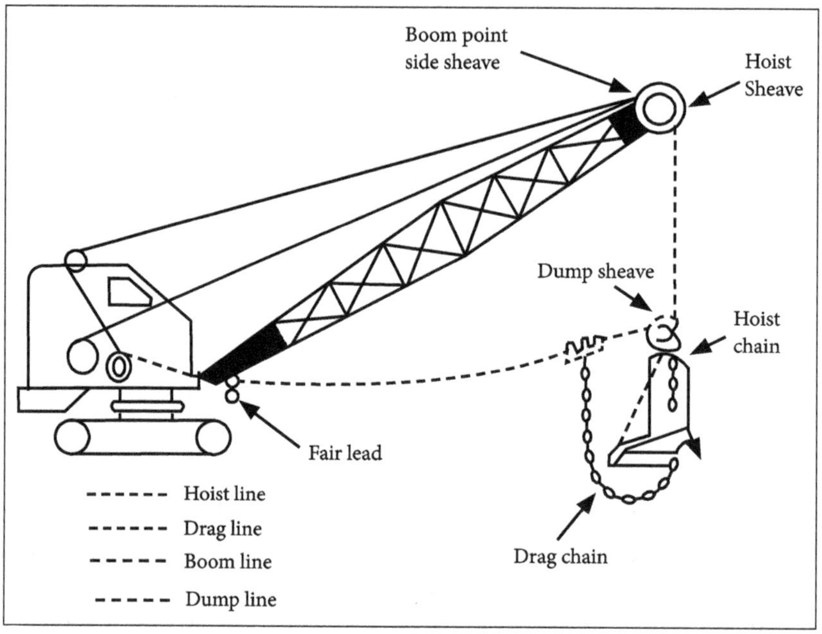

Dragline.

A dragline can dig and dump over longer distances than an ordinary shovel can do.

Because of its long boom, the machine need not be seated close to the pit. It can handle digging of underwater excavation while stationing the machine on a firm soil away from the location of the pit. This can dig below the track level and can handle soft materials comfortably.

(iv) Clamshell

It gets the name as its bucket resembles to a clam which is a shell-fish with a hinged double shell. Digging is done as in a dragline and soon after the bucket is filled it works like a crane. It has a crane boom with a specially designed bucket attached to it at the upper end through cables. The bucket is hinged at top and has either sharp edge or the teeth at the bottom.

During digging process the bucket is lowered with shells open. When reaching the surface to be dug it makes a good contact with it. The weight of the bucket helps the sharp-edge or the teeth at the bottom of the bucket to dig into the surface of material, thereby filling the bucket. Once the bucket is filled the shells close.

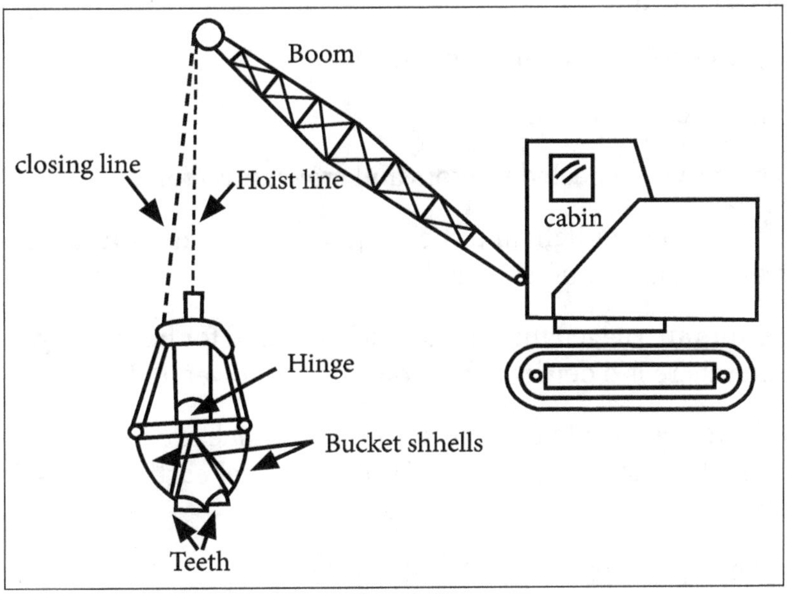

Clamshell

The bucket is then hoisted and swung to the position where dumping has to be done and the contents are dropped. Buckets may be of light or heavy weight and used depending on the type.

3. Bulldozers

It is an important equipment on a construction project. It is basically a scraping and pushing unit. However, it is multipurpose equipment which can be used for different purposes with some modifications. Accordingly they are called as angle dozer, tilt dozer, tree dozer and push dozer.

Angle dozer pushes its load at an angle (nearly 25°) to the direction of travel. This is helpful:

- When the material has to be pushed down the slope on hill work.
- Where a long windrow has to be made during the travel of the dozer.

Tilt dozer is used to start excavating a ditch or a trench or for excavation in hard ground. For this the blade is required to be tilted by raising one end up to 25 cm above the other. Push dozer is used to push the scraper unit after digging and also during loading operation using the pusher plate. Tree dozer is used to up-root and removes trees.

4. Tractors

A tractor is a multi-purpose machine. It comes in varied types as light model to heavy model. The light model is used for agricultural or small haulage purposes. Heavy model equipped with several special rigs are used for earth moving work. This is an important piece of equipment which is indispensable on all important projects. Two principal applications of tractors are:

- Clearing and excavating machinery.
- Hauling and conveying machinery.

There are two types, viz., wheeled tractors and crawler tractors.

The wheeled type is used for light and speedy jobs. As regards to its applicability it falls between the crawler tractor and the truck.

The crawler tractors are rugged machines which are used for heavy duty work. It is used particularly when there is a demand for more tractor power and speed of movement.

Wheeled units have been made to work on the jobs which are intended for the crawler type. Wheeled tractors are not available for all practical earth moving jobs including ripping and dozing.

Following are the important qualities of wheeled tractors:

- These can travel faster than crawler tractors and can be used where it has to be moved to long distance with high speed.
- These have wheel steering control as in automobile vehicles and are easily measured.
- These can be self-driven over long distances.
- These tractors need less skill of operation and less maintenance cost.
- These are comparatively less in cost.

Important Qualities of Crawler Tractors

Following are the important qualities of crawler tractors:

- These are more compact and powerful which can handle heavy jobs of hauling and digging.
- These when used for long distances are done with trailers.
- These are not liable to slip over very smooth or loose footings when increased power is applied to the wheels.
- These are when fitted with special shoes can be moved on pavements or on tarred roads.

5. Motor Graders

A grader is primarily a device for loading or finishing earthwork. It is also used for mixing gravel, making windrows and trimming slopes. There are two types of graders, viz., Towed and Motorized.

The towed type is of small size with a tractor. The controls are in general manual and sometimes a small petrol engine is fitted on the framework of the grader to operate the controls. It is operated by separate operator. The power to operate the controls is supplied through a quill shaft passing through hollow clutch and transmission shafts.

Irrespective of the movement of the machine, the quill shaft is made to rotate whenever the engine is running. The versatility of the machine is increased by the addition of these tools. The blade is thick and like a blade of dozer and can swivel through 180°.

The scarified is the tool for loosening hard soil and may be mounted in the front or in the rear of the machine. The bulldozer and the snow-plow (a V-shaped blade) are front-end attachments fitted to the grader for pushing loads or clearing snow.

The elevator attachment enables the grader to pick up the material cut by the blade and drop it over an inclined belt conveyor which transmits it into carrier unit. In order to compact or smoothen the graded surface, roller may be attached to the rear of the grader.

6. Scrapers

Scrapers are the devices to scrap the ground to load the material, to transport to the required distance, to dump at the intended site, to spread the dumped material over the required area, to attain the desired thickness and to return back to do the next cycle. Scrapers are designed to dig, load, haul, dump and spread. As a scraper does a multiple works it is also called as carry all.

The process of unloading by the scraper is called as dumping and spreading operation. In this operation, the bowl is positioned so as to spread the material. To have an even spreading, a partial opening of the apron in the stage of unloading will be helpful. In the case of wet and sticky material, the apron should be raised and lowered repeatedly.

This helps the material behind it to get loosened and material drops out of the bowl by moving the ejector forward, the remaining material is pushed out at a uniform rate. After the dumping is complete, the tail gate is fully retarted, the apron is dropped and the bowl is raised to the transporting position. Using the cutting edge final finishing is done.

Following are the types of scrapers:

- Towed Type.
- Self-propelled.
- Self-loading.
- Tandem Scraper.
- Twin Engine Scraper.
- Multi-engine Multi-bowl Scraper.

A scraper consists of the following parts:

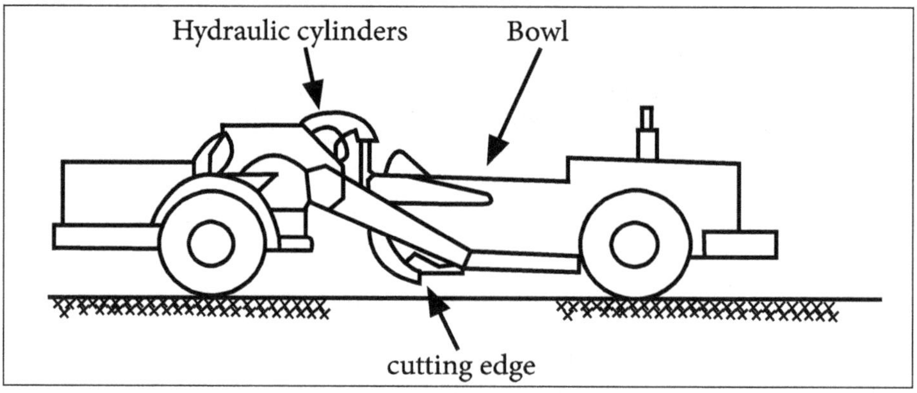

Scraper

- Bowl,
- Apron,
- Ejector,
- Hydraulic System,
- Digging or Loading,
- Transporting.

7. Loaders

It is a machine in which a bucket is attached to the arms and capable of being raised, lowered and dumped through mechanical or hydraulic controls. A very common type is one in which the loader is having the bucket in the front which is known as front-end loader.

Loaders are self-propelled and versatile equipment which are mounted either on crawler or wheel-type running gear. These are fitted with front mounted general purpose bucket.

These are operated through hydraulic rams with which they dig, scoop, lift, transport, carry, dump or load into hauling units, bins, hoppers, conveyors and stockpiles. With the additional parts added to the front and rear mounted attachments these can doze, scrap, grab, forklift, trench, grade, ditch, rip, clamp and winch.

Loaders are used for handling coal, sugar, sand, salt, stone, etc., to earth moving and digging works in quarries.

(i) Crawler Loaders

These are low pressure types which are generally preferred for digging and loading jobs of poor ground and rock and sharp stony ground.

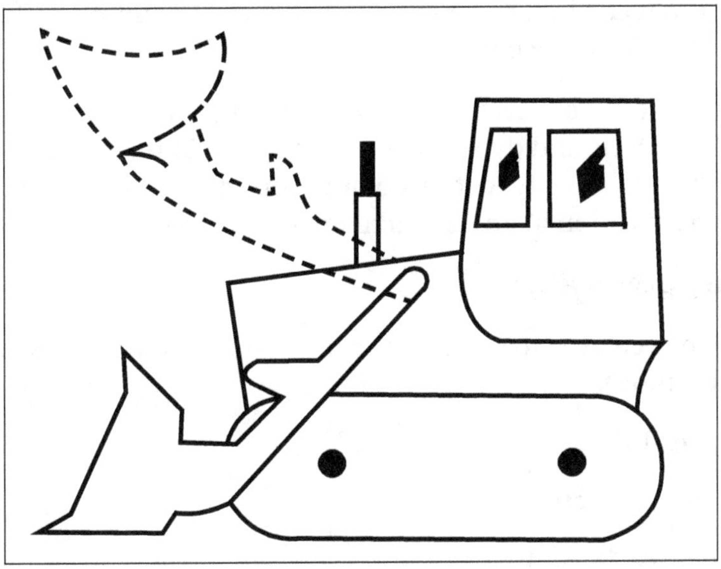

Crawler loader.

(ii) Wheeled Loaders

These are generally four wheel drive. Two-wheel drives are preferred for handling light jobs on good ground conditions.

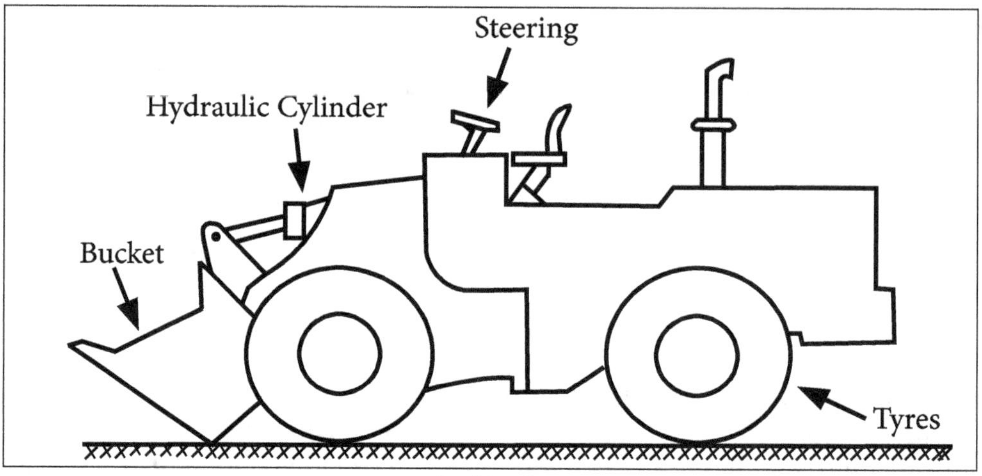

Wheel loader.

Equipment for Compaction

1. Compaction

It is the process by which particles of materials are packed closely together by reducing air void content. Compaction is achieved by mechanical means. Application of mechanical forces may be static or dynamic.

Static forces are produced by a roller or compactor and the number of times, the roller has moved over the soil is called as "passes".

Dynamic energy, produce a vibratory or tamping effect on the soil. To achieve the required density both static and dynamic forces are used in the compactors. Apart from producing the required density, compaction produces a reasonable surface finish according to the layer position in an embankment or in a road.

2. Compaction Specification

Two methods have been used to specify the compaction requirements in the compacted soil layers (Foster, 1962).

The two types are called as:

- Performance-type specification.
- Work-type specification.

(i) Performance - Type

In the performance-type specification, the compaction requirement is stated in terms of the physical properties of the compacted layer. One of the physical properties adopted is the percentage of maximum dry density obtained in a standard compaction test. It is applicable for cohesive soils.

(ii) Work - Type

In the work-type specification, the type of equipment, the lift thickness, the moisture content and the amount of work required to obtain the necessary density are specified.

3. Road Structure

A road consists of several layers with each layer having a specific function. Different layers of a road or runway of an airport are shown in figures below.

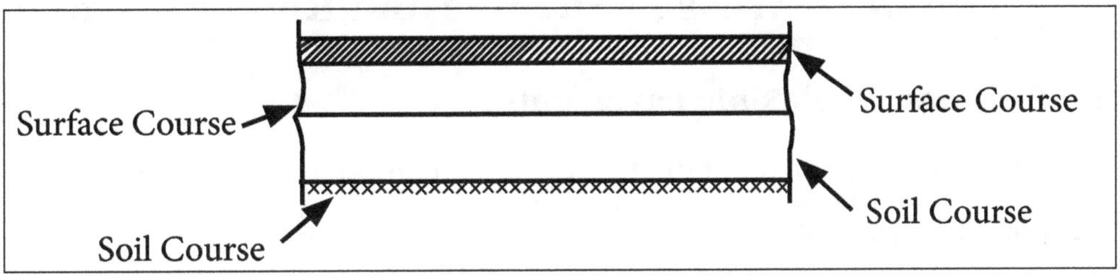

Flexible Road structure.

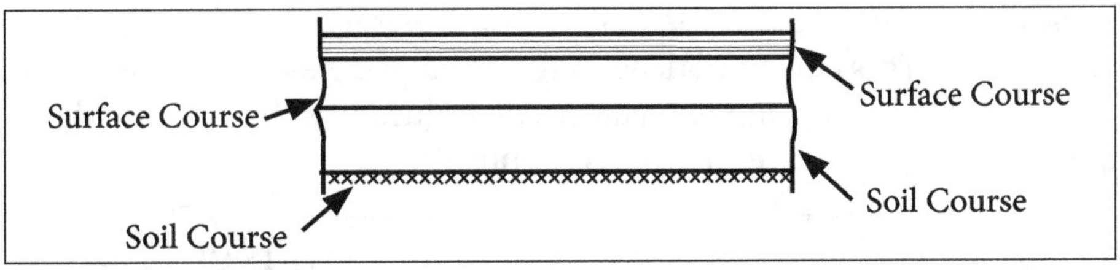

Rigid Road structure.

4. Embankment Structure

Embankments are constructed for highway or railway to lay on it in order to have an elevated position for some reason or the other. Major portion of the embankment is constructed in successive layers which are compacted using rollers or compacter. Each layer is called as lift thickness:

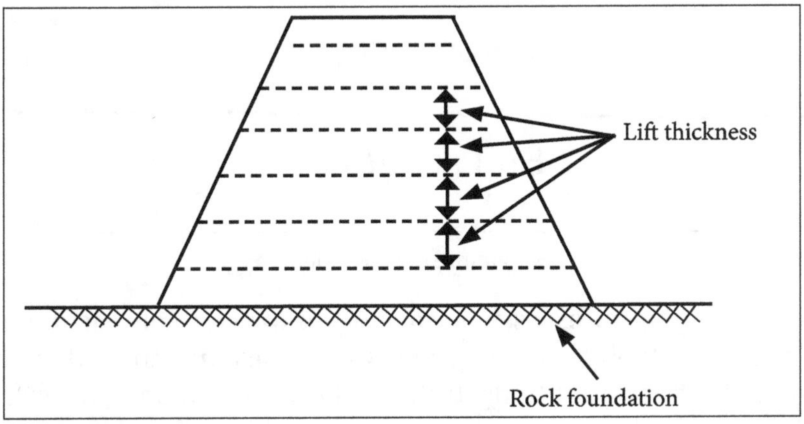

Earth embankment.

5

Evaluation and Maintenance of Pavements

5.1 Pavement Distress in Flexible and Rigid Pavements

Causes of Distress in Flexible Pavements

It may be attributed to settlement of any one of the components of the pavement.

Failure in Subgrade Soil

The main cause for failure of flexible pavement is the excessive settlement of the subgrade of soil. Figure shows the settlement of the subgrade and the effects on the base and wearing course. This can be noticed in the form of excessive undulations or waves and corrugations on the pavement surface. The failure of subgrade may be attributed to inadequate stability and excessive stress application.

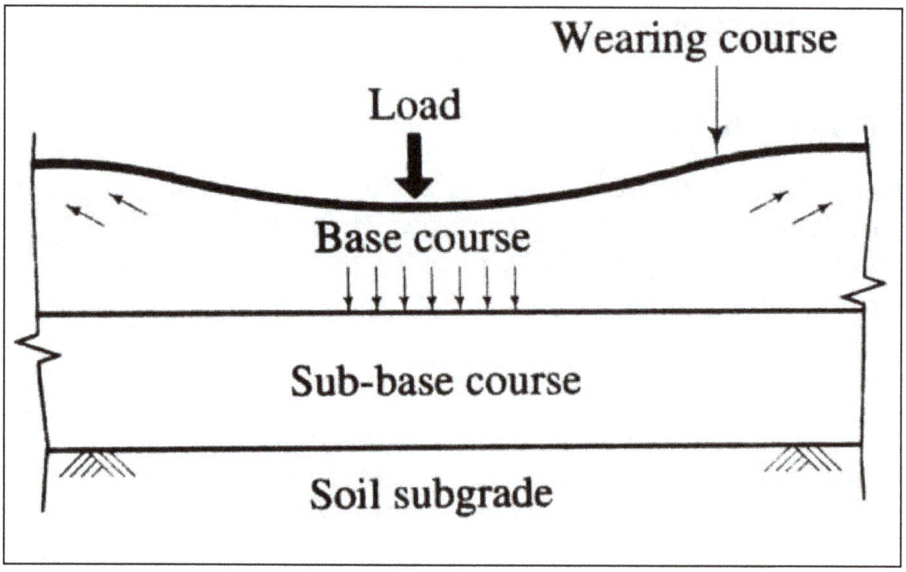

Failure of subgrade.

Subgrade stability is due to excessive moisture, improper compaction and inherent weakness of the soil. Stability of subgrade is the resistance to settlement under the short and long term stress conditions. If the load transferred on to the subgrade due to wheel loads is less then the deformation of the subgrade will be less and will be in the elastic range.

Continued stress application causes the deformation to be plastic and unrecoverable resulting in subsidence of the subgrade. Figure shows the elastic and plastic deformations of a pavement.

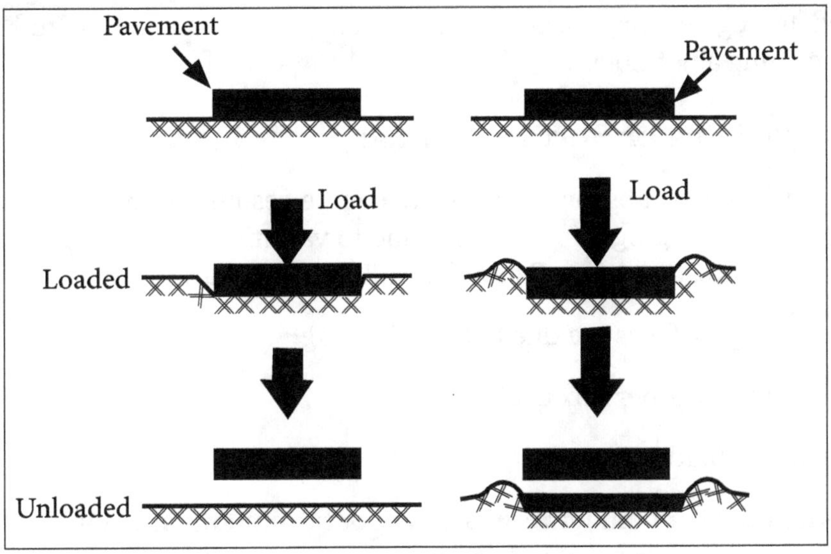

(a) Elastic Deformation (b) Plastic Deformation
Deformation of Pavement.

Failures of Sub-base or Base Course

Sub-base and base courses may settle due to internal readjustment of aggregates and movement of wheel loads. May result in loosening of the compacted layer accounting for disturbance and loss of binding action. Thus deformation would occur.

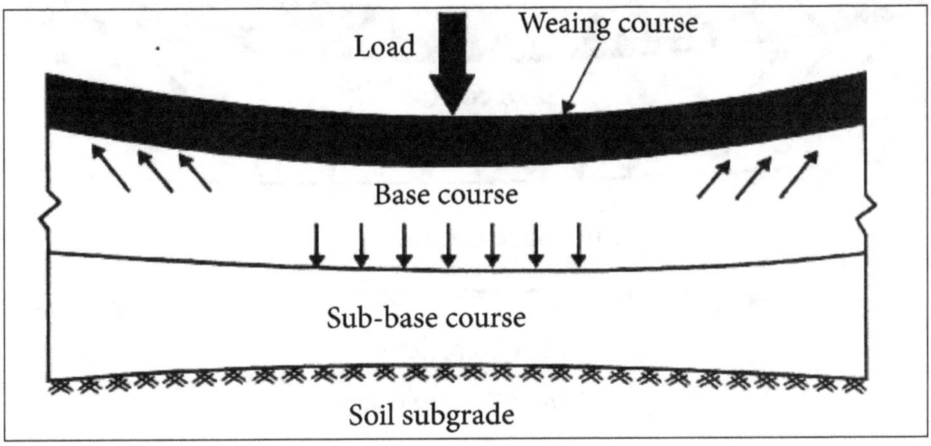

Failure of Base Course.

Inadequate thickness or poor mix of base or sub-base courses may lack stability. Improper quality control during construction of sub-base or base course may cause settlement.

Loss of base course materials is also possible when either the base course is not covered properly or has completely worn out. Due to moving vehicles, a sort of suction is caused which leads to removal of binding material in WBM base.

The exposed aggregates may form dust due to abrasion and attrition and loose stone aggregates forming a pot hole.

Failure of Wearing or Surface Course

Inadequate thickness or absence of wearing course results in damage of base course immediately. The damaging effect is more due to variation in climatic conditions particularly in rain or frost weather. Example: Bullock carts traffic.

Failures of wearing surfaces are due to the following:

- Inferior or improper mix design.
- Inadequate binder cement.
- Inferior quality of binder results in a poor bituminous surfacing.

Volatilization and oxidation of binders also make the bituminous surfacing brittle. Because of this chemical action, cracking of the pavement occurs which allows seepage of rain water to harm the underlying layers.

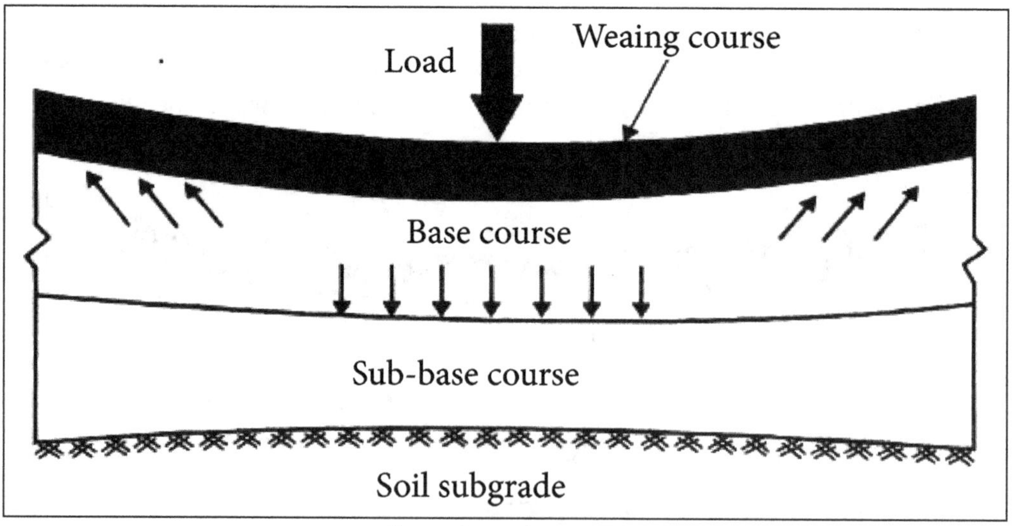

Failure of Wearing Course.

Failure Due to Inferior Materials and Construction

Many failures are attributed to the use of inferior materials in the pavement. Some materials may appear to satisfy initially but may show rapid deterioration. Continued weather change further deteriorates all the layers resulting in cracking.

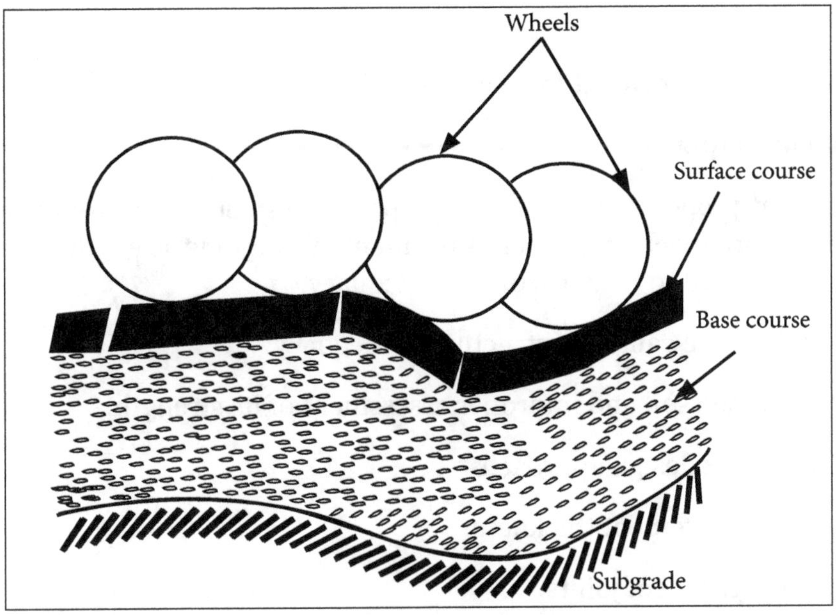

Failure due to improper compaction of subgrade and pavement layer.

Improper construction of layers and inadequate quality control may yield failure of one or all the layers of the pavement. Figure shows failure of base course, surface course and subgrade because of improper compaction.

Causes of Distress of Rigid Pavements

Distress or failure in cement concrete pavements may be attributed to three factors as detailed below:

- Deficiencies in pavement materials.
- Structural inadequacy.
- Improper construction and maintenance.

Deficiencies in Pavement Materials

Failures or distress in cement concrete pavements are due to improper selection of materials and their quality:

Following are the chief causes.

- Soft aggregates.
- Dirty aggregates with silt and clay.
- Low quality joint filler.
- Poor sealer material.

- Poor quality steel.
- Improper use of cement for the specific region.

Structural Inadequacy

Non-collection of proper traffic data, improper assessment of future traffic demand, improper design approach, improper assessment of subgrade strength, climatic condition, etc. are attributed to the structural inadequacy.

Following are the main causes of structure inadequacy:

- Poor subgrade soil and improper assessment of its strength.
- Improper mix design approach.
- Inadequate pavement thickness.
- Incorrect spacing of joints.
- Incorrect design of load transfer devices.
- Absence of longitudinal hinge joints.
- Long length of slab.
- Non-existence of temperature steel.
- Deep foundation movements.

Improper Construction and Maintenance

Many of the failures or distress is due to the following factors:

- Poor workmanship in pavement and joint construction.
- Poor surface finish.
- Improper and insufficient curing.
- Use of concrete mixes which are wet.
- Over finishing at the edges.
- Improper alignment of dowel bar and lubrication of bars.
- Infiltration of incompressible materials (like sand) into the joint.
- Unplanned annual maintenance.
- Special maintenance after adverse weather conditions.

5.2 Pavement Management Systems

It is a planning tool used to aid pavement management decisions. PMS software programs model future pavement deterioration due to traffic and weather. It recommends maintenance and repairs to the road's pavement based on the type and age of the pavement and various measures of existing pavement quality.

Measurements can be made by persons on the ground, visually from a moving vehicle or using automated sensors mounted to a vehicle. PMS software helps the user create composite pavement quality rankings based on pavement quality measures on roads or road sections.

Recommendations are usually biased towards preventive maintenance. Typical tasks performed by pavement management systems include inventory pavement conditions, identifying good, fair and poor pavements.

It assigns important ratings for road segments based on traffic volumes, road functional class and community demand. It schedule maintenance of good roads to keep them in good condition and schedule repairs of poor and fair pavements.

Research has shown that it is far less expensive to keep a road in good condition than it is to repair it once it has deteriorated. Pavement management systems place the priority on preventive maintenance of roads in good condition rather than reconstructing roads in poor condition.

In terms of lifetime cost and long term pavement conditions, this will result in better system performance. Agencies that concentrate on restoring their bad roads often find that by the time they've repaired them all, the roads that were in good condition have deteriorated.

The State of California was among the first to adopt a (PMS) in 1979. First PMS was based in a mainframe computer and contained provisions for an extensive database. It can be used to determine long-term maintenance funding requirements and examine the consequences on network condition if insufficient funding is available.

5.3 Factors Pertaining to Pavement

Pavement Evaluation

A study of various factors pertaining to pavement, such as subgrade support, pavement composition and its thickness, traffic loading and environmental condition, is referred to as pavement evaluation. Aim of pavement evaluation is to assess as to whether and to what extent the pavement fulfills the design requirements.

Based on the evaluation, the maintenance and strengthening works could be planned. Study is used to investigate the structural adequacy of pavements and also the requirements for providing safe and comfortable traffic operations.

Although several methods are used, the four methods categorized below are the basic ones:

- Structural Evaluation of pavements.
- Evaluation of pavement surface condition.
- Present serviceability index.
- Visual Rating.

Roughness of Surface

Apart from different aspects, the riding quality of a pavement is a function of the specifications adopted for the surfacing. Roughness is one among them. A measure of the pavement performance can be obtained by monitoring its roughness.

Smooth riding surface brings about many advantages, viz., higher speeds, less fuel consumption, less wear and tear of tyre, less consumption of spare parts, riding comfort and safety. In performance evaluation, rigidity of surface plays a vital role.

Present Serviceability Index

A rating system was developed by AASHO involving the measurement of permanent deformation, riding quality and the extent of cracking and patching. Rating is well-known by the term Present Serviceability Index (PSI). It is the most widely used pavement rating measure.

Following equations give the value of PSI for flexible and rigid pavements,

1. Flexible Pavements

$$ISI = 5.03 - 1.91\log(1+SV) - 1.38(RD)^2 - 0.01\sqrt{C+P}$$

2. Rigid Pavements

$$ISI = 5.41 - 1.80\log(1+SV) - 0.09\sqrt{C+P}$$

Where,

SV - Slope variance giving an index of the longitudinal profile.

RD - Rut depth under a 4 ft. straight edge.

C - Cracking distress in terms of lineal feet of cracks per 1000 sq.ft., area.

P - Patched area in sq.ft. per 1000 sq.ft., area.

Skid Resistance

Though smooth surface has various advantages it is dangerous to traffic, especially when the surface is wet and the vehicles move fast. It is normal to construct reasonably skid-resistance surfaces. Passage of vehicles polishes the aggregates and excess Bitumen tends to fatten the surface and render it slippery.

Hence it is essential to evaluate the skin resistance of the surface at periodic intervals to ensure that the roughness level has not fallen to dangerous low levels. It is accomplished by periodic measurement of skin-resistance. Skin resistance is measured using Towed fifth wheel pump integrator, Car-mounted roughness measuring device, etc.

Structural Evaluation of Pavements

Plate bearing test can be conducted for both flexible and rigid pavements to assess the structural capacity. Assessment may be made by the load carried at a specified deflection at a place or by the amount of deflection at a specified load on the plate. Performance of a flexible pavement is closely related to the elastic deformation under loads or its rebound deformation.

Under the existing conditions of a pavement, the deflection of pavement under the design wheel loads serves as an index to carry traffic loads. Benkelman Beam is the most commonly used method.

Before conducting the Benkelman Beam measurements, a rating survey of the road is performed by dividing the road into homogeneous sections of approximately similar serviceability.

Deflection Measurement

It occurs due to various factors such as subgrade soil type, moisture content and compaction level of subgrade, pavement thickness and drainage conditions, pavement surface temperature and wheel load.

Measured using a suitable equipment such as Benkelman Beam, Lacroix Deflectograph, Dynaflect and Deflectome.

5.4 Strengthening of Existing Pavements

A highway is expected to have adequate stability to withstand the design traffic under prevailing climatic and subgrade conditions. If the design load and the anticipated traffic are maintained, the planned maintenance will be fruitful.

If the pavement is to support increased wheel loads and load repetitions, the pavements cannot be maintained properly but undergo large distress. Only solution to manage the increased traffic is either to direct the traffic on some adjacent roads or to strengthen the existing pavements.

Strengthening may be done by providing additional thickness of pavement provided the subgrade is strong enough. Construction of one or more layers over the existing pavement is called overlay. If the distress of the road is excess, construction of overlay will not help. The road has to be reconstructed.

Types of Overlays

An overlay may be either flexible or rigid. Flexible overlays are generally bituminous surfaces. It can be constructed over existing flexible pavements or existing concrete pavements. A bond breaker comprising of a granular base is used between a concrete base pavement and overlay. Deflection is also reduced to some tolerable level.

Rigid overlays may consist of plain, simply reinforced or continuously reinforced concrete pavements. Provision of a bond breaker may or may not be necessary. Following are the combination of overlays:

Existing pavement	Overlay
Cement concrete	Cement concrete
Cement concrete	Bituminous
Bituminous or flexible	Cement concrete
Bituminous or flexible	Flexible or Bituminous

Choice of overlay depends on various factors including total thickness of overlay required, Wheel load, Subgrade strength, etc.

5.5 Types of Maintenance

Maintenance operations of highways fall under three categories noted below:

- Routine maintenance,
- Special repairs,
- Resurfacing.

1. Routine Maintenance

All types of pavements are exposed to climatic changes and adverse weather conditions, evidently these would wear out. To avoid further deterioration and to reduce further heavy expenses, it is mandatory to maintain the highway in time.

The following works are to be maintained routinely:

- Up keeping of carriage way.
- Shoulders.
- Subgrade.
- Side drain clearance for silt accumulation.
- Highway geometric and traffic control.
- Ancillary works like culverts, bridges, etc.

2. Special Repairs

Adverse weather conditions and other causes may demand special repairs of the pavement which would have been damaged. Need may arise for special repair of some sub-surface drainage system and cross drainage structure.

3. Resurfacing

It is arise due to the expiry of the life of the pavement or surface being heavily damaged.

5.6 Highway Project Formulation

A new highway project comprises of design and construction either at network of new roads or a road link. There may also be projects which require redesign and re-alignment of existing roads of upgrading the geometric design standards. A new highway project formulation may be divided into the following stages:

- Selection of route and survey of possible alignment of roads.
- Finalization of highway alignment.
- Collection of materials and testing of subgrade soil and other construction materials.
- Design aspects.
- Construction stages.

Selection of Route and Surveys

Based on the points to be connected, different routes are tried. Selection of route, the requirement of alignment and the geological, typographical and other features of the locality are kept in mind. Surveys to be conducted for selection of route comprises of the following steps:

1. Map Study

This is the preliminary stage of the project.

At this stage the help of available maps of the area are studied.

2. Reconnaissance Survey

Tentative alignments are primarily looked into keeping in various factors. To decide the correct alignment a reconnaissance survey is conducted. Following are the points to be considered during the reconnaissance survey:

- The alignment should be such that necessary cutting and banking are avoided.
- Obstructions to be faced along the alignment are looked into.
- Gradients and length of curves to be provided in the alignment.
- The alignment should avoid any cross-drainage work.
- The alignment should not be taken through agricultural land completely.
- The type of soil along the route should be looked into.
- The sources of construction materials along the route to be assessed.
- The type of terrain to be noted.
- The magnetic bearing of the alignment are measured by compass.
- Nature of ground on either side of the alignment and the highest flood level are noted.

3. Preliminary Survey

The preliminary survey or the preliminary location survey of the highway project is carried out with the following main objectives:

- Various alternate alignments.
- Estimation of the quantity of earthwork materials and other construction aspects.
- Compare different proposals.

Following surveys are conducted during the preliminary survey:

- Primary traverse.
- Topographical surveys.

- Levelling work.
- Collecting hydrological data.
- Soil surveys.
- Establishment of permanent benchmarks by running a fly levelling.
- Intersection of canals, of rivers is to be properly located.
- Drawings are prepared following route survey, longitudinal and cross- sections, mass hall diagram, contour map of the strip and design of curves.
- At the office, the cost estimate of different works such as land compensation, flyovers, culverts, bridges, earthwork, etc., are prepared.

Finalization of Highway Alignment

After the preliminary surveys, the final alignment is decided and is first located on the field by establishing the center line. Next the detailed survey should be carried out and the geometric design details are made. The detailed survey involves the following:

- Centre line is marked at intervals of 30 m.
- Fixing temporary benchmarks along the route for every 300 m.
- Cross section details are taken for 30 m on either side of the central line.
- All details of cross drainage works are taken.
- Topographical details are taken.
- Detailed soil survey is carried out.
- Total land width required is marked by pillars at regular intervals.
- Tangent points and intersection points of the curves are properly marked by pillars.

Collection of Materials and Testing

Soil samples collected during the survey along the selected route are tested in the laboratory in order to design the pavement and the design of embankments and cuttings. Selected construction materials from borrow areas are also tested for the required properties.

Mix-design tests of materials are conducted in the laboratory to explore the possibility of using them for the pavement component layers. The possibility of using low-cost construction material like soil aggregate mixes, soft aggregates, stabilized soil and concrete mixes, in the sub-base or base course layers of pavement should also be fully explored.

Design Aspects

After the survey and investigation are over, a new project involves in the following design steps:

- Identifying the class of road to be designed and the terrain over which it has to pass.
- Design speed selection appropriate to the class of road and the terrain.
- Selection of cross sectional elements keeping in view the future traffic requirements.
- Selection of appropriate geometric design standard suitable to the class of road, terrain and design speed.
- Design of alignment and profile.
- Alignment design of embankments and deep cuts.
- Choosing of appropriate pavement based on the need and material availability.
- Design of pavement.
- Design of surface and subsurface drainage system and erosion control.
- Design of cross-drainage structures.
- Design of road junctions.
- Estimation of various quantities of items and the total cost.
- Estimation of road use cost.
- Estimation of maintenance cost.
- Comparing various alternatives and selecting the optimal solution.
- For the selected route, preparation of final drawings and estimates.
- Preparation of final project report.

Construction Stages

The construction of the road may be divided into two stages, viz.,

- Earth work.

- Pavement construction.

The earthwork comprises of excavation and construction of embankments.

During the excavation process for highway cuts, protection of earth slopes and construction of drainage works are taken care of and construction of highway embankment by rolled-fill method is preferred. Compaction of soil in layers under controlled moisture and density are done using suitable layers. In situations of high embankments special checks such as stability of the embankment foundation and slopes, settlement, etc. are to be done.

Permissions

All chapters in this book are published with permission under the Creative Commons Attribution Share Alike License or equivalent. Every chapter published in this book has been scrutinized by our experts. Their significance has been extensively debated. The topics covered herein carry significant information for a comprehensive understanding. They may even be implemented as practical applications or may be referred to as a beginning point for further studies.

We would like to thank the editorial team for lending their expertise to make the book truly unique. They have played a crucial role in the development of this book. Without their invaluable contributions this book wouldn't have been possible. They have made vital efforts to compile up to date information on the varied aspects of this subject to make this book a valuable addition to the collection of many professionals and students.

This book was conceptualized with the vision of imparting up-to-date and integrated information in this field. To ensure the same, a matchless editorial board was set up. Every individual on the board went through rigorous rounds of assessment to prove their worth. After which they invested a large part of their time researching and compiling the most relevant data for our readers.

The editorial board has been involved in producing this book since its inception. They have spent rigorous hours researching and exploring the diverse topics which have resulted in the successful publishing of this book. They have passed on their knowledge of decades through this book. To expedite this challenging task, the publisher supported the team at every step. A small team of assistant editors was also appointed to further simplify the editing procedure and attain best results for the readers.

Apart from the editorial board, the designing team has also invested a significant amount of their time in understanding the subject and creating the most relevant covers. They scrutinized every image to scout for the most suitable representation of the subject and create an appropriate cover for the book.

The publishing team has been an ardent support to the editorial, designing and production team. Their endless efforts to recruit the best for this project, has resulted in the accomplishment of this book. They are a veteran in the field of academics and their pool of knowledge is as vast as their experience in printing. Their expertise and guidance has proved useful at every step. Their uncompromising quality standards have made this book an exceptional effort. Their encouragement from time to time has been an inspiration for everyone.

The publisher and the editorial board hope that this book will prove to be a valuable piece of knowledge for students, practitioners and scholars across the globe.

Index

Q

R

S

T

U

V

W

www.ingramcontent.com/pod-product-compliance
Lightning Source LLC
LaVergne TN
LVHW080601210926
848485LV00003B/90